IN DEDICATION

To Richard, who loved being '<u>in</u> the shadow' as much as anyone else. He touched the lives of many through his love of the starry skies and his exceptional abilities as a teacher.

Richard A. Sweetsir
1944-1995

Observe Eclipses
Second Edition

by

Dr. Michael D. Reynolds
and
Richard A. Sweetsir

Cover — The 22 December 1870 total solar eclipse expedition party of Lord Lindsay in Cadiz, Spain. Lindsay also lead expeditions to observe the 1874 transit of Venus and a solar eclipse expedition to India in 1871, where he photographed the corona. Lord Lindsay is standing at the eyepiece of the telescope. *Photograph courtesy of the Royal Astronomical Society.*

Published by:
The Astronomical League
Science Service Building
1719 N Street, NW
Washington D. C. 20036

Copyright © 1995 and 1979
by
Michael D. Reynolds
and
Richard A. Sweetsir

All rights reserved. Except for brief passages quoted in a review, no part of this book may be reproduced by any mechanical, photographic, or electronic process, nor may it be stored in any information retrieval system, transmitted, or otherwise copied for public or private use, without the written permission of the authors. Requests for permission or further information should be addressed to the Astronomical League.

First Published 1979

Printed in the United States of America

ISBN 1-886336-10-5

The Astronomical League OBSERVE Series

Observe Eclipses
A Guide to the Messier Objects
Observe the Herschel Objects
ALPO Mars Observers Handbook
Observe and Understand the Sun

Other Astronomical League Publications

Astronomy Teachers Handbook
Math for Amateur Astronomers

The following publishers have generously given permission to use quotations or work from copyrighted works: from *The Sky: a user's guide* Copyright © 1991 by Cambridge University Press. Reprinted by permission of Cambridge University Press. From *Gerald's Game* by Stephen King. Copyright © 1992 by Stephen King. Used by permission of Viking Penguin, a division of Penguin Books USA Inc. From *Peanuts* Copyright © by UFS, Inc. and Charles Schulz. Reprinted by permission of UFS, Inc. From *Minolta Master 8-778* Copyright © 1994 by the Minolta Corporation. Reprinted by permission of the Minolta Corporation. Photograph of Richard Sweetsir taken by Scott Robinson, *Jacksonville Journal*; *Florida Times–Union* photograph Copyright © 1994 Florida Publishing Company.

Table of Contents

Chapter	1	Historical Perspective	1
		Lunar and solar eclipses, American eclipses, literary references, contemporary experiences.	
Chapter	2	Eclipse Dynamics	7
		Eclipses defined, the earth's shadow; total, partial and penumbral lunar eclipses; apparent size of the moon and sun; total, annular, annular–total, and partial solar eclipses; orbital motions of earth and moon, the moon's orbital plane; eclipse seasons and frequency; visibility, duration and magnitudes of eclipses, the saros, the inex series, the Metonic cycle.	
Chapter	3	Eclipse Observing and Vision Safety	13
		Solar observing, the sun at totality, lunar observing, what's safe and what isn't, projection method; pinhole, mirror and optical projectors; viewing filters, rear– and front–mounted filters, other approaches, unsafe methods reemphasized.	
Chapter	4	Expedition Planning	18
		Commercial ventures, independent expeditions, ECLIPSE planning method (establish, chronicle, localize, instrument, practice, sustain, evaluate).	
Chapter	5	Solar Eclipse Observing – The Partial Phases	23
		Contact timings, sunspot contact timings, limb irregularities, photographic projects.	
Chapter	6	Solar Eclipse Observing – Environmental Studies	25
		Pinhole camera images, diminishing light, wildlife behaviors, meteorological observations, weather satellite imagery, photographic projects.	
Chapter	7	Solar Eclipse Observing – Shadow Bands	27
		Explanations, appearance, equipment, procedures, observations from totality's edge, photography.	
Chapter	8	Solar Eclipse Observing – The Lunar Shadow and Sky Darkness	29
		Appearance, procedures, photography.	
Chapter	9	Solar Eclipse Observing – The Diamond Ring and Baily's Beads	31
		Explanations, procedures, photography, viewing dangers.	
Chapter	10	Solar Eclipse Observing – Second and Third Contacts, Flash Spectrum, Chromosphere, and Prominences	34
		Contact timings, flash spectrum, chromosphere, prominences, photography, viewing dangers.	
Chapter	11	Solar Eclipse Observing – The Corona	36
		Shape and extent, intensity variations, streamers, photography, viewing dangers.	
Chapter	12	Solar Eclipse Observing – The Sky at Totality	38
		Planets, variable stars, comets, meteors and fireballs, artificial satellites, faintest star, zodiacal light, gegenschein, photography.	
Chapter	13	Lunar Eclipse Observing – The Penumbral Eclipse	40
		Detection timings, shadow position, umbral separation; shading and color variations; photoelectric photometry and intensity variations; other observations, photography.	
Chapter	14	Lunar Eclipse Observing – The Partial Phases	42
		Program management, contact timings, crater contact timings, umbral characteristics, apparent magnitude estimates, lunar transient phenomena, photoelectric photometry, photography.	
Chapter	15	Lunar Eclipse Observing – Totality	51
		Contact timings, general color and uniformity, eclipse luminosity, apparent magnitude, lunar transient phenomena, stellar total and grazing occultations, meteors, variable stars, artificial satellites, photoelectric photometry, photography.	
Chapter	16	Eclipses of Other Types	54
		Planetary transits, eclipses of planetary satellites, occultations, eclipsing binary stars, spacecraft events.	
Chapter	17	Eclipse Photography	59
		Film, photographic equipment, focusing, photographing a solar eclipse, photography of other solar eclipse phenomena, photographing a lunar eclipse, photography of other lunar eclipse phenomena, archiving your photographs, summary.	
Chapter	18	Electronic Eclipses – Video and CCD Imaging	66
		Video imaging, equipment, mounting the video camera, charge–coupled devices, eclipse observations using electronic imaging, summary.	
Appendix	1	Recommended Books	72
Appendix	2	Recommended Periodicals and Software	73
Appendix	3	Glossary of Selected Terms	75
Appendix	4	Astronomical Organizations.	77
Appendix	5	List of Suppliers	78
Appendix	6	Solar Eclipses, 1900 – 2025	79
Appendix	7	Lunar Eclipses, 1900 – 2025	83
Appendix	8	Astronomical League Solar Eclipse Report Form	86
Appendix	9	Eclipse Observer's Log	87
Appendix	10	A.L.P.O. Satellite Eclipse Report Form	88
Appendix	11	Request Form for NASA Solar Eclipse Bulletins / NASA Solar Eclipse Bulletins on the Internet	89
Index			91

Preface and Acknowledgements

The popularity of eclipse astronomy has not waned since the first edition of **Observe Eclipses**, published in 1979. In fact the opposite is true—eclipse chasers travel world-wide in search of those fleeting minutes of being in the shadow of the moon. Observers also travel distances to set up at the limits of annular eclipses, to find clear skies for a total lunar eclipse and to observe transits of Mercury.

We considered a simple revision when asked about a second edition of **Observe Eclipses**. However so much has changed in the 15 years since the first edition that a complete rewrite was in order. A chapter on historical and contemporary eclipse experiences has been added as well as chapters on eclipse photography and "Electronic Eclipses"—video and CCD eclipse astronomy. The appendices are greatly expanded and will provide the observer—neophyte as well as the veteran eclipse chaser—a wealth of information and resources.

We are especially pleased with the contributions to the second edition of **Observe Eclipses**. The first chapter reflects this aspect, as does the center color plate section, photographs and several sections of the appendices. We are grateful to Jay Anderson, Steve Edberg, Fred Espenak, David Garcia, David Levy, Ludwig Meier, Jose Olivarez, Jay Pasachoff, Dorothy Pillmore, and Carter Roberts for their contributions to Chapter 1's contemporary perspectives. Some of their anecdotes are quite humorous and bring to life the adventure of eclipse chasing.

The authors would also like to acknowledge Jay Anderson and Don Trombino for their contributions to Chapter 4, Norm Sperling for Chapter 9, and Jay Anderson, Fred Espenak and John Westfall for their contributions to the appendices.

The photos, CCD images and videos contributed to this edition represent some of the best eclipse photography. Our sincere gratitude to Jay Anderson, Doug Berger, Steve Edberg, James Engelbrecht, Milt Hays, Alan Gorski, Conrad Jung, Mike Kazmierczak, J.S. Korintus, Mike Martinez, Derald Nye, Don Parker, Jeremy Reynolds, Carter Roberts, Ewell Schirmer, Volkmar Schorcht, Don Trombino, and John Westfall. Lick Observatory, the National Aeronautics and Space Administration and the Royal Astronomical Society also contributed several photographs.

We also wish to acknowledge the photography of Peter Calabrese, Francis Graham, Rusty Harvin, Larry Loper, Aimee Reynolds and Karl Simmons. We would also like to acknowledge the commercial firms Carina Software and Tom Mathis, Minolta Camera Corporation and Kosuke Sasaki, Roger Tuthill Inc. and Roger Tuthill, Sun Spotter and Daniel R. Janosik Sr., and Thousand Oaks Optical for supplying illustrations and photographs.

This second edition is also rich in artwork and illustrations. Art and astronomy go hand in hand, so it is appropriate that such rich illustrations are a part of this new edition. Vina Loper's beautiful depiction of a cruise ship under the shadow of the moon adorned the cover of the first edition; the reader can again find it in this edition. Jeanne Pusateri's artwork can be found in several chapters, as can Jeremy Reynolds' depiction of Zeus in Chapter 1 and George Murphy's artwork depicting the Apollo–Soyuz Test Project artificial eclipse experiment in Chapter 16. Carroll Hebbel's cartoons about expeditions are a perfect addition to the chapter on expedition planning as is Fred Espenak's cartoon in Chapter 1. And everyone recognizes the magic of Charles Schulz and *Peanuts*; two comics from his series on the 20 July 1963 total solar eclipse have been included. Scientific and technical illustrations include graphics of future eclipses by NASA Goddard Space Flight Center's Fred Espenak, eclipse dynamics illustrations by David Frantz, Carter Roberts' photographic exposure graphs and John Westfall's illustrations of craters for lunar eclipse timings and CCD field coverage.

David Frantz is responsible for the professional layout of this edition. David's patience with us ("could you please change this to that; now change it back") is gratefully acknowledged.

The authors wish to acknowledge the following individuals for reviewing the text and providing other invaluable assistance and advice: Stephen Bieda, Jr., Ted Cox, Francis Graham, Rusty Harvin, Etta Heber, Bob Menezes, Theresa Nelson, Jose Olivarez, Don Parker, Carter Roberts, Norm Sperling, Ruth Sweetsir, Larry Toy and John Westfall. Roy Bishop of the Royal Astronomical Society of Canada, Ron Maddison, and Peter Hingley, Royal Astronomical Society Librarian, were very helpful in confirming specific eclipse information and Diane Reid and Natasha Cooper of United Media were helpful in researching *Peanuts*. Any errors or omissions are entirely the fault of the authors, not of these individuals.

We would also like to thank our families for their patience while enduring this project. Writing and layout is enjoyable to many but can be frustrating to the "writing widowed family."

Finally the authors wish to acknowledge the Astronomical League and especially Rollin P. Van Zandt. Van's passing was a great loss to all of us; we fondly remember his assistance with the first edition of **Observe Eclipses** and prior to his death his enthusiasm, support for and contributions to this second edition.

We hope you as readers and especially as observers enjoy the second edition of **Observe Eclipses** as much as we enjoyed writing it! We wish you the clearest of skies 'in the shadow.'

Michael D. Reynolds, Ph.D.
Richard A. Sweetsir

And it shall come to pass in that day, saith the Lord God, that I will cause the sun to go down at noon, and I will darken the earth in the clear day:
The Bible, Amos VIII:9

Historical Perspective

Introduction. Scholars attribute this Old–Testament scriptural reference, from the Book of Amos, to an eclipse of the sun which Assyrian records date to 15 (7[*]) June 763 B.C., but there are accounts of lunar and solar eclipses earlier in antiquity.

Ancient eclipse accounts are found throughout the literature and history of most major civilizations. Correctly attributing such accounts to actual events is a difficult and challenging task once the exclusive territory of scholars. With the proliferation of home computers and astronomy software, however, even this bastion of academia has become fertile ground for amateur hobbyists.

A plethora of books and articles have been written about the archaeoastronomers of Western and Northern Europe and the nature of their enigmatic structures. Were artifacts such as those at Stonehenge (England) and Carnac (France) ancient observatories, primitive celestial calculators, calendars or religious sites? These and similar structures, which have been dated variously from 4000 B.C. to 1500 B.C., continue to generate debates over their significance to eclipse astronomy.

Early peoples perceived eclipses of the sun and moon as portents of evil. They were seen as battles in the sky between the two celestial objects, or as the devouring of one or the other by dragons, coyotes, snakes, jaguars or other beasts. It was reasonable for these peoples to assume that noise might frighten away the transgressor and return the sun or moon to the sky; since arrangements for suitably noisy demonstrations took some time, the ability to correctly predict eclipses was of great importance.

Lunar eclipses. The earliest account of a lunar eclipse may be of one visible from the Middle East that some scholars date as far back as 3450 B.C. Babylonian astrologers recorded observations of an eclipse of the moon believed to have taken place in 2283 B.C. Others date the oldest recorded lunar eclipse with greater certainty to 1361 B.C., probably the total of 15 (3) February; accounts of this event were found transcribed on ancient oracle bones in China.

Ptolemy records that lunar eclipses were observed in 721 B.C. and 720 B.C., while the Athenian historian Thucydides relates that a lunar eclipse on 28 (23) August 413 B.C. convinced the Athenian general Nicias to postpone evacuating his army from Sicily for nearly a month; by then, Sparta's forces managed to block his escape route, destroy his fleet and bring about a major Athenian defeat in the Peloponnesian War.

Inscriptions on a clay tablet, part of an ancient Chaldean Astronomical Almanac, tell of lunar and solar eclipses that took place in the same month (dates are Chaldean):

> *On the first Mercury rises.*
> *On the third the Equinox.*
> *Night of the 15th. 40 minutes after sunset, an eclipse of the moon begins.*
> *On the 28th. occurs an eclipse of the sun.*

Some sources date these two eclipses to 9 (4) October and 23 (18) October 425 B.C.

Solar eclipses. The Chinese may have been predicting eclipses as early as 2800 B.C., relying on an 18–year cycle of similar eclipses. Chinese legend describes the fates of two astronomers, Hsi and Ho, who were beheaded for neglecting their duties and failing to warn their ruler of a solar eclipse. Efforts to date this eclipse, from what little data are available, have placed it as early as 2165 B.C. and as late as 1948 B.C. Some sources cite an eclipse of 22 October 2136 B.C., but none occurred on that Julian date, although a partial eclipse was visible from China on 11 October (23 September). Other sources claim the eclipse of the legend more likely occurred sometime after 1952 B.C.

A clay tablet excavated at ancient Ugarit, within the borders of modern–day Syria, contains an account of a solar

---

[*] For events prior to 1582, the original Julian calendar date is given, followed (in parentheses) by its modern Gregorian calendar counterpart.

Figure 1-1 Chinese astronomers Hsi and Ho, unsuccessful in predicting an eclipse, prepare to meet their fate! *Illustration by Jeanne Pusateri.*

eclipse believed to be that of 3 May (21 April) 1375 B.C. The earliest confirmed record of a solar eclipse may also provide the first description of the sun's corona. As with the lunar eclipse of 1361 B.C., ancient Chinese oracle bones provide the documentation. Although the historian Breasted cites 1217 B.C., modern calculations show no solar eclipses visible from China that year, although a partial was visible north of China across Siberia on 25 (14) May; a better candidate might be the annular of 6 June (26 May) 1218 B.C. which was partial in China. Liu Chao-yang suggests the bones date to 1353, 1307, 1302 or 1281 B.C.

Babylonian astronomers also discovered that eclipses of the sun were periodic, which is to say, they recurred at specific intervals of time. An Ionian statesman by the name of Thales used Babylonian observations obtained in his travels to predict the solar eclipse of 28 (22) May 585 B.C., gaining him great fame in the process. This eclipse was credited by the historian Herodotus with ending a six-year war between the Medes and the Lydians.

Thucydides describes an annular solar eclipse that occurred on 3 August (29 July) 431 B.C. during the Peloponnesian War:

> ...the sun assumed the shape of a crescent and became full again, and during the eclipse some stars became visible.

Chaldean astronomers began keeping extensive records of their astronomical observations as early as 568 B.C., and continued the practice without interruption for over 360 years. Shortly before 500 B.C., the Chaldean astronomer Nabu-rimannu used these observations to compile tables that, among other notable achievements, exactly dated solar and lunar eclipses.

Nabu-rimannu was followed more than 100 years later by Kidinnu, who used the then-much-larger body of Chaldean data to compile even more accurate tables. The later Greeks, notably Meton, used these observations and calculations to independently discover periodic cycles of eclipses.

A description of the total solar eclipse of 15 (10) August 310 B.C. by Agathocles during his voyage from Syracuse to Africa has allowed for a calculation of his probable route as being north of Sicily.

American eclipses. The first lunar eclipse observed by Europeans in the new world was most likely that of the night of 29 February (10 March) 1504. Christopher Columbus resolved a standoff over food supplies with the natives of Jamaica by threatening to extinguish the moon; Columbus obtained the much-needed supplies when his anticipated lunar eclipse took place.

The first total solar eclipse observations of record from the British Colonies were made on 24 June 1778 by Philadelphia astronomer David Rittenhouse; the path of this eclipse ran from southern California to New England.

The first official U. S. total eclipse expedition was launched in the midst of the Revolutionary War. Led by the Reverend Samuel Williams, Hollisian Professor of Mathematics and Natural Philosophy at Harvard, the expedition arranged for safe passage through British lines to Penobscot, Maine (then part of Massachusetts) to observe the eclipse of 27 October 1780. Unfortunately, due to either British restrictions, faulty maps, or a calculation error, the expedition set up camp just outside the path of totality!

Even from the edge of the eclipse path, however, Williams was successful in observing, describing and sketching an interesting eclipse phenomenon. Williams wrote: *The sun's limb became so small as to appear like a circular thread or rather like a very fine horn. Both the ends lost their acuteness and seemed to break off in the form of small drops or stars some of which were round and others of an oblong figure. They would separate to a small distance, some would appear to run together again and then diminish until the whole disappeared.*

The phenomenon Williams described bears the name "Baily's beads" for the English observer Francis Baily who noted them 56 years later, during the annular eclipse of 15 May 1836.

Other American expeditions were launched for the total eclipses of 1806 (Massachusetts), 1834 (South Carolina), 1869 (Kentucky), 1878 (Wyoming), and 1889 (California). The annular solar eclipse of 17 September 1811 was observed by Thomas Jefferson at Monticello two-and-a-half years after he left office as the third President of the United States. Jefferson, who had a deep interest in astronomy, made careful timings of the moon's contacts by telescope.

The 7 August 1869 eclipse expedition, led by Joseph Winlock, third director of Harvard Observatory, is notable for the participation of George and Alvan G. Clark, sons of Alvan Clark, famed telescope maker and founder of the Alvan Clark and Sons optical firm, and John Adams Whipple, daguerreotypist and pioneer astrophotographer. The team obtained one of the most spectacular photographs of the solar corona to that date.

Uncertainty over the nature of the symbols used to depict various astronomical objects and events have made it difficult to identify pictographic records of eclipses among the artifacts of the American Indians. One such record left by the Dakota tribe may depict the total solar eclipse of 7 August 1869.

Photograph 1-2 Total solar eclipse of 1 January 1889 at Cloverdale, California. Two second exposure by Charles Burckhalter, Chabot Observatory. Burckhalter used a 10.5–inch f/8 Newtonian reflector.

Literary references. Lunar eclipses appear in literature with far less frequency than solar eclipses. An early literary reference to a lunar eclipse dates to one seen from Athens on 9 (4) October 425 B.C.; the same verse refers to the annular solar eclipse of 21 (16) March 424 B.C., which was partial from Athens. They are noted in **The Clouds**, a comedy written by the Greek playwright Aristophanes in 423 B.C.:

> And the Moon in haste eclipsed her,
> and the Sun in anger swore
> He would curl his wick within him
> and give light to you no more
> **Aristophanese, Chorus of Clouds**

There is a reference in Homer's Greek classic, the **Odyssey**, which has been attributed by Plutarch and Eustathius to a solar eclipse that was total around Ithaca on 16 (5) April 1178 B.C.

> ...and the Sun has perished
> out of heaven,
> and an evil mist hovers over all.
> **Homer**

This surviving fragment of a Greek poem by Archilochus is believed to refer to the total eclipse of the sun that occurred in mid–morning on 6 April (30 March) 648 B.C. at Thasos.

> Zeus, the father of the
> Olympic Gods, turned
> mid–day into night, hiding the light
> of the dazzling Sun;
> and sore fear came upon men.
> **Archilochus**

Another Greek poet, Pindar, wrote of the solar eclipse of 30 (25) April 463 B.C., which appeared nearly total at Thebes, while Cicero reports that the Roman poet Quintus Ennius described the solar eclipse of 21 (16) June 400 B.C.

Plutarch wrote of a solar eclipse, ...*which, beginning just after noon, showed us plainly many stars in all parts of the heavens, and produced a chill in the temperature like that of twilight.* Some authorities identify this eclipse as the annular–total of 20 (18) March 71 A.D., while others label it a work of pure fiction.

Many centuries later, the essence of early human attitudes toward solar eclipses was captured vividly by Milton in his 1667 epic **Paradise Lost**:

> The sun...
> In dim eclipse, disastrous twilight sheds
> On half the nations, and with fear of change
> Perplexes monarchs.
> **Milton**

In **Samson Agonistes** (1671), Milton provides this vivid description of a total solar eclipse:

> O dark, dark, amid the blaze of noon,
> Irrecoverably dark, total eclipse
> Without all hope of day!
> **Milton**

Not all human eclipse beliefs focus on evil, however. Romance is at the root of Tahitian lore, where the sun and moon hide one another as they make love; their offspring are the stars. A more benign view of eclipses is also shared by some Arctic peoples who believe the sun and moon leave their normal positions during eclipses to make certain that everything is going well on the earth.

Some authors have exhibited a striking lack of familiarity with eclipse dynamics. Sir Henry Rider Haggard's original 1885 edition of his famous novel **King Solomon's Mines** depicts a full moon, a solar eclipse, and another full moon occurring on successive days, an impossible sequence of events. His second edition corrected the error by replacing the solar eclipse with a significantly more plausible lunar eclipse.

Thomas Hardy's 1903 poem, **At a Lunar Eclipse**, employs vivid geometric allusions to capture the essence of the celestial spectacle:

> At a Lunar Eclipse
> Thy shadow, Earth, from Pole to Central Sea,
> Now steals along upon the Moon's meek shine
> In even monochrome and curving line
> Of imperturbable serenity.
>
> How shall I like such sun–cast symmetry
> With the torn troubled form I know as thine,
> That profile, placid as a brow divine,
> With continents of moil and misery?
>
> And can immense mortality but throw
> So small a shade, and Heaven's high human scheme
> Be hemmed within the coasts yon arc implies?
>
> Is such a stellar gauge of earthly show,
> Nation at war with nation, brains that teem,
> Heroes, and women fairer than the skies?
> **Thomas Hardy**

More recently, the popular novelist Stephen King set two of his works, **Gerald's Game** (1992) and **Dolores Claiborne** (1993) partially against a backdrop of the 20 July 1963 total solar eclipse in Maine. In **Gerald's Game**, King writes: *Above her, a furnace of strange light glowed fiercely around the dark circle hanging in the indigo sky.... She was faintly aware that the owl was still calling, and that the crickets had been fooled into beginning their evensongs two or three hours early. An afterimage floated in front of her eyes like a round black tattoo surrounded by an irregular halo of green fire....*

Contemporary experiences. Rapid global transportation has made it possible for many more astronomers, professionals and amateurs alike, to place themselves inside eclipse shadows. In addition to conducting traditional scientific observations, many choose to express themselves in more artistic ways, through poetry, prose and art. In recognition of the contributions of these modern–day Miltons, the selections that follow are offered as representative samples.

Noted astronomer, lecturer, author and comet discoverer David Levy, in his book **The Sky: a user's guide**, describes the total lunar eclipse of 30 December 1963 as an exception to the generally held view that lunar eclipses are far less inviting than their solar equivalents:

A sharp cold front had rushed through, leaving the Montreal sky clear and bitterly cold. The eclipse began with a barely detectable penumbral shadow; we could not detect any darkening at all until this shadow was almost half way across the face of the moon. During the last few minutes before the arrival of the earth's umbra... the moon was significantly darker.

The umbra attacked the moon like an onrushing army. Our satellite was literally disappearing; where was the dull red glow we had expected? The moon had shrunk to a thin crescent which narrowed and disappeared altogether.

The eclipse was now total, and later we were to learn it was one of the darkest on record. Through a telescope we were able to see the moon's dim outline, but we needed a star chart to find it.

Amateur astronomer and eclipse enthusiast Dorothy Pillmore submits this account of another dark lunar eclipse:

Members of the Northern Colorado Astronomical Society assembled at Horsetooth Dam high above Fort Collins to observe the 9 December 1992 total lunar eclipse. At our locality, the moon would rise already totally eclipsed. We were lined up along the dam with clear view from north to south for our telescopes, cameras and binoculars.

Time passed for moonrise and we saw—nothing! A white 'spot' finally appeared about 15 degrees above the horizon and widened into a sliver of a crescent. The eclipse was so dark we simply hadn't seen it! The moon was a dark silvery gray with a bluish cast. It was too dark to time craters emerging from the earth's shadow and was by far the darkest lunar eclipse I've seen.

Fred Espenak, astrophysicist with N.A.S.A.'s Goddard Space Flight Center, shares this anecdotal account of an African expedition threatened by inclement weather:

Thousands of people traveled to Kenya for the 16 February 1980 total solar eclipse. My four–observer group decided the best weather strategy was to go into the bush where we would have plenty of maneuverability if conditions took a turn for the worse; the remaining 90 or so eclipse chasers at our hotel elected to observe the eclipse from the comfort of the hotel.

With five minutes left until second contact the cloud cover had increased to 95%; with two minutes left we made the decision to run!

Grabbing only one camera with a long lens, I jumped into the front seat of our minibus to direct our driver, Ali. The others grabbed the tape recorder, quartz clock and one of the telescopes. Still struggling with the telescope under which he was literally trapped, one of my group made several vain attempts to close the bus door and end the trail of soda bottles, lens caps and empty film boxes we left in our wake.

Rumbling down the road at 100 kilometers per hour, I watched the diamond ring form while hanging out the front window. After timing the contact we traveled another 100 meters before commanding Ali to stop. As I opened the front door and jumped out, panic exploded in the back of the bus; nobody could get the rear door open! In quick succession the others dove over the telescopes and seats to emerge out the front door and remove our equipment. We furiously began our observations and photography; Ali simply smiled as he watched our frantic flurry of activity.

The corona itself was very bright with a larger streamer to the northwest. Even through the clouds, a lot of structure was visible during this sunspot–maximum eclipse. Through my camera viewfinder, I spotted a large prominence at 12 o'clock. To the southeast the sky was bright orange as I looked out of the edge of the moon's shadow.

Time passed as did our hole in the clouds. Twenty seconds before third contact, the corona was lost from view as the cloud bank covered the sun. Noting no break in the clouds, I grabbed the tape recorder, quartz clock and tripod and took off sprinting down the road after the hole, two of my group right behind me. We caught up to the hole and watched third contact on the run.

As I strolled back to our minibus on an adrenaline high I realized that I had never fully appreciated the term 'eclipse chaser' until now! And back at the hotel, our comrades enjoyed better weather during the eclipse, but they missed a most memorable adventure!

Astronomer Jay Pasachoff offers this composition as a personal biographical addendum, first published in **Scientific American**:

Light Verse

Higgledy–piggledy,
Jay Myron Pasachoff
Williams astronomer,
Dabbles in rhyme.

Solar eclipses and
Radio telescopes
Keep him contented
The rest of the time.

Jay Pasachoff

Eclipse-chaser Carter Roberts relates some of his experiences:

A group of us went to Colombia for the 12 October 1977 eclipse despite the poor weather prospects. We selected a site near the center line where totality would last 53 seconds. Going up the dirt road leading to the site, the bus got stuck in the mud twice when the driver shifted gears in the middle of each mudhole. We finally selected a different site where totality would last 35 seconds. Three minutes before totality a cloud blocked the sun and stayed until 10 minutes after.

Near Roy, Montana, for the 26 February 1979 eclipse, I ducked when I thought I was going to be hit by the shadow! The lighting was so strange, I also saw a distant mountain range appear to float. This was one of the best eclipses I have been to.

I went to Bratsk, Siberia, for the 31 July 1981 eclipse. The day before the eclipse we were flown 600 miles west to Kemerovo, north of the path of totality, (and) eclipse morning were taken to the old airport at Leninsk-Kuznetski where we observed the eclipse in perfectly clear skies. This was the only eclipse at which I did not see clouds in front of the sun at some time during the eclipse.

Environment Canada Meteorologist Jay Anderson offers these insights he acquired as forecaster for the 18 March 1988 **Sky & Telescope** magazine total eclipse cruise.

*We were aboard the **Golden Odyssey**, north of Borneo, sailing northeastward down the center line the day before the eclipse. Steve Edberg, Leif Robinson and I were responsible for putting the ship in the correct place and we agreed to meet in the wheelhouse at 4 am to discuss our prospects.*

The weather had been pretty good—lots of thin cirrus, but that's usually hard to escape in equatorial regions. The National Weather Service, Honolulu, told me I would have a weak easterly wave approaching our area. In the tropics, an easterly wave is a small weather disturbance which brings showers and thundershowers—but the area in front of them is usually cloud free. Honolulu indicated that it was very weak; we might even be able to observe the eclipse while in the midst of it.

I couldn't sleep that night—after all, I was the tour's meteorologist, and everything would depend on my advice. Finally I gave up and went on deck.

Lightning could be seen in the distance ahead—the easterly wave we were expecting. I had two choices—punch through it and hope we would run clear of the cloud by eclipse time, or turn back to the southwest, and try to keep in front of the disturbance.

'What do you want me to do?' said the Captain, after I had outlined our situation to the three of them.

*I looked at Steve—he was the man who handed out the latitude and longitude figures, and gave the Captain the eclipse coordinates. I looked at Leif. It was his, or at least **Sky & Telescope's**, tour group. 'I don't know,' they echoed.*

'Well, tell me something,' said the Captain.

'Turn back and head southwest,' I offered, with more authority than I really felt. I was looking at the lightning ahead, wondering if we could really run through it. Besides, if we stayed in front we would be blessed with virtually cloud-free skies, provided this disturbance followed the usual pattern.

The easterly wave behaved just as the textbooks said. An hour after the eclipse, just as lunch was beginning, the wave passed with its retinue of cloudy buildups, showers and thundershowers, and delighted the diners with a spidery waterspout just off the starboard side. What a wonderful eclipse that was!

Reflecting on the 11 July 1991 total solar eclipse, Texas amateur David Garcia reports: *I was so engrossed... I almost forgot my cameras!*

Jose Olivarez, director of the Omnisphere and Science Center in Wichita, Kansas, is accustomed to receiving telephone inquiries regarding astronomical events; the annular eclipse of 10 May 1994 was no exception:

One of the more interesting calls was from a woman who informed me she was keeping her dog indoors. The woman then told me that she was afraid the dog's eyes would be damaged if the dog looked at the eclipse!

Photograph 1-3 Observers are reminded to take all precautions for safe solar viewing; filters and naked-eye protection are essential. This observer gives new meaning to the term "sun dog."

Dr. Ludwig Meier, R&D Scientist at Carl Zeiss Jena, relates some interesting experiences from a couple of solar eclipse expeditions:

We were interested in measuring inner and outer corona photographs taken through a specially-designed camera. The camera had a system of four—63 mm diameter lenses; three matched with polarizers and one non-polarized lens. Later the coronal brightness could be calibrated.

Our first attempt was the 2 October 1959 eclipse in the Canary Islands where we were promised clear weather prospects of 95%... it was cloudy. Our next expedition was to Brac Island, Yugoslavia for the 15 February 1961 where weather prospects were a dismal 25% chance of clear skies... it was clear!

Fishing was one Brac Island occupation; the fishermen would go at night in a small boat with a fork-like spear and use a light to attract fish. We were stunned when the fishermen went fishing during totality!

Astronomer Stephen J. Edberg of N.A.S.A.'s Jet Propulsion Laboratory, shares the following recollections:

Every eclipse I've gone to has yielded a distinct memory, and most have taught a lesson as well.

You always hear stories of solar astronomers forgetting to remove a lens cap during totality. It happened to me in 1991 during the grand eclipse in Baja California, though the loss ended up not mattering. And another significant mistake I made yielded success instead of failure!

My telescope assembly for that eclipse consisted of a Newtonian flash spectrograph, a 2300 mm f/29 large format refracting system, a 485 mm f/5.5 refractor, and a 17 mm f/3.5 wide angle zodiacal light camera. About 15 minutes before totality a single cloud condensed near the sun, and then promptly evaporated. I watched this, while around the same time I went through the process of pulling lens caps and white cloth covers (for keeping cameras cool) from the instruments.

Inexplicably, I forgot to pull the cap and cover from the wide angle camera. With the onset of second contact I went into action, successfully recording flash spectra, the corona in a full exposure series, and the prominences and inner corona with the long focal length instrument through totality and third contact. I also remembered to open the shutter in the wide angle camera, a little later after second contact. I even remembered to close it after third contact, a problem had the sky been better. But it turned out the sky was so bright during totality that it would have washed out the zodiacal light anyway, so there was no real loss and I have a good human–fallibility story to tell, and it was made obvious to the world on a PBS special on the eclipse.

The other mistake ended more than a decade of frustration. I have been trying to record the flash spectrum for many years. Incorrect exposure has been the main problem, though chromatic aberration in some systems was an accessory to the lack of success. Thus, this slitless Newtonian system was my latest attempt, using color negative film (with its wide latitude) to try to avoid the earlier bugaboos. I avoided them all right, but in an unexpected manner: it turned out that I accidentally chose the non–blazed (less bright) first order spectrum generated by the diffraction grating, instead of the blazed side that I intended when I collimated this instrument with the others. The exposures were perfect for the second and third contact flash spectra—SUCCESS at last!

Finally, the co–authors of **Observe Eclipses** are not without their own unique experiences chasing eclipses.

Richard Sweetsir flew to Dar Es Salaam, Tanzania for the total solar eclipse of 23 October 1976, but his plans to fly on to Zanzibar Island for the early morning event met with unforeseen complications:

A last–minute decision by Tanzanian president Julius Nyerere to observe the eclipse from the island prompted his security forces to refuse passage to day–before travelers, even those already ticketed. Appeals to East African Airways, armed Tanzanian security forces, and the U.S. Embassy fell on deaf ears; we were forced to observe from the mainland beach at Bagamoyo, northwest of the capital. From our site, one thick convection cumulus cloud, formed by warm air currents above Zanzibar, blocked the sun throughout the eclipse; observers on the island, including President Nyerere, had an unobstructed view of totality!

Mike Reynolds shares this amusing encounter with a Canadian meteorologist just prior to the 26 February 1979 total solar eclipse:

As we had done prior to several previous total solar eclipses we visited the local weather station both the day prior to and the morning of the eclipse. When we asked the meteorologist at the Winnipeg, Manitoba station why the sudden clearing the morning of the 26th, he simply replied, 'It was a trough aloft, eh?'

Photograph 1-4 Contemporary total solar eclipse watchers can chase the moon's shadow from the air. Observers are preparing for the 31 July 1981 eclipse near the Hawaiian Islands. Note the seats have been removed to allow easy access to the jetliner's windows. *Photograph taken by Peter Calabrese.*

References

Alter, Dinsmore and Cleminshaw, Clarence. **Pictorial Astronomy**. New York: Thomas Y. Crowell Company, 1956.

Bartlett, John. **Familiar Quotations, 14th ed**. Boston: Little, Brown and Company, 1968.

Bible, The Holy Authorized King James Version. New York: Harper & Brothers.

Breasted, James Henry. **Ancient Times: A History of the Early World, 2nd ed**. Boston: Ginn and Co., 1944.

Encyclopaedia Britannica, 15th ed. Chicago: Encyclopaedia Britannica, Inc., 1980.

Gomme, A. W. **A Historical Commentary on Thucydides**. London: Oxford University Press, 1956.

Hutchins, Robert Maynard, ed. **Great Books of the Western World, vol. #5** (The Plays of Aristophanes, translated into English Verse by Benjamin Bickley Rogers). Chicago: Encyclopaedia Britannica, Inc., 1952.

Jones, Bessie Zaban, and Boyd, Lyle Gifford. **The Harvard College Observatory: The First Four Directorships, 1839-1919**. Cambridge, MA: The Belknap Press of Harvard University Press, 1971.

King, Stephen. **Gerald's Game**. New York: Viking, 1992.

Krupp, E. C., ed. **In Search of Ancient Astronomies**. New York: Doubleday & Company, Inc., 1978.

Link, F. **Eclipse Phenomena in Astronomy**. New York: Springer-Verlag Inc., 1969.

Levy, David, H. **The Sky: a user's guide**. Cambridge: Cambridge University Press, 1991.

Mitchell, S. A. **Eclipses of the Sun**. New York: Columbia University Press, 1923.

Moore, Patrick. **The Guinness Book of Astronomy Facts & Feats**. London: Guinness Superlatives Limited, 1979.

Raphael, Henry. **Thomas Jefferson, Astronomer: Leaflet No. 174 - Aug., 1943**. San Francisco: Astronomical Society of the Pacific.

Schove, D. Justin. **Chronology of Eclipses and Comets AD1 to 1000**. Woodbridge, Suffolk, Great Britain: The Boydell Press, 1984.

Eclipse Dynamics

Eclipses defined. An *eclipse* is an apparent dimming or extinction of light coming from one heavenly body caused by another body. In the case of a *lunar eclipse*, the full moon enters the shadow of the earth. A *solar eclipse* occurs when the new moon passes in front of the sun.

A lunar eclipse may be *total*, *partial* or *penumbral*; an understanding of these three types requires some knowledge of the nature of the earth's shadow. A solar eclipse may be *total*, *annular*, *annular–total*, or *partial*, the first three collectively are referred to as *central eclipses*; knowledge of the changing apparent sizes of the sun and moon in the sky is helpful in understanding these solar eclipse types.

Finally, familiarity with the orbital characteristics of the earth and moon, especially the plane of the moon's orbit with respect to the earth, is essential for an understanding of the dynamics of both lunar and solar eclipses.

The earth's shadow. The central cone of the earth's shadow, called the *umbra*, stretches out into space a distance which averages 1,379,000 km (857,000 mi). At the distance of the moon's orbit, this shadow measures some 9,200 km (5,700 mi) in diameter, more than 2.6 times the diameter of the moon itself. From within this cone, a hypothetical observer on the moon would see the earth completely obscure (totally eclipse) the sun.

In addition, there is a secondary shadow surrounding the umbra, called the *penumbra*, which, at the moon's distance, is nearly twice as wide, averaging 16,000 km (10,000 mi) in diameter. Within this outer shadow, a lunar observer would see the earth graze or partially block the sun, but never completely obscure it. These dynamics allow for three kinds of lunar eclipses, the total, the partial and the penumbral.

Total lunar eclipse. A total eclipse of the moon takes place when the moon passes centrally through the earth's umbra (and, of course, penumbra). The earth's shadow may be seen crossing the moon's disk during the partial stages of a total eclipse until the entire disk is immersed in shadow at the onset of totality. It should be

Figure 2-1 Lunar Eclipses, showing the relative size of the moon with the earth's umbra and penumbra. *Illustration by David Frantz.*

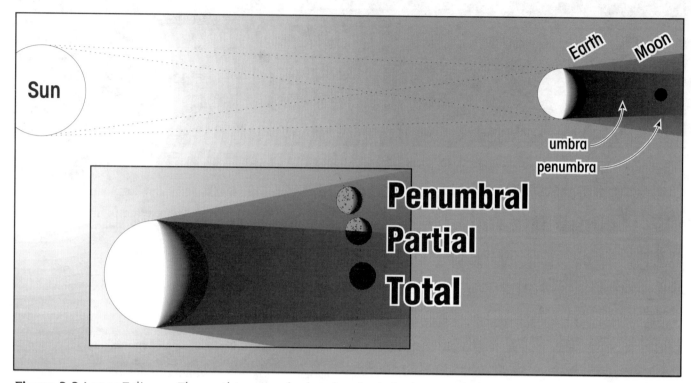

Figure 2-2 Lunar Eclipses. The earth's penumbral and umbral shadow in relation to penumbral, partial and total lunar eclipses. (Due to space limitations, the sun is shown proportionally closer to the earth than it is in reality, therefore the shadow sizes and shapes are not according to scale.) *Illustration by David Frantz.*

noted, however, that the moon does not completely disappear from view because the earth's atmosphere refracts (bends) some sunlight from the earth's sunlit side, and the moon reflects this dim light with a reddish–brown glow even when totally eclipsed.

Partial lunar eclipse. A partial eclipse of the moon is similar to the partial stages of a total eclipse, except the moon is never completely obscured; the moon's umbral passage is not central. Our hypothetical moon–based observers fortunate enough to be within the umbra would see the earth totally eclipse the sun, while observers outside the umbra would see only a partial solar eclipse.

Penumbral lunar eclipse. When the moon enters the earth's penumbral shadow but misses the umbra altogether, a penumbral eclipse of the moon occurs, the least spectacular of eclipse phenomena. The earth's penumbral shadow is so light that the moon must be within 1120 km (700 mi) of the umbra for it to produce any detectable change in the moon's appearance, and even then it presents only a subtle reddish color shift. Moon–based observers would see the earth partially eclipse the sun.

Apparent size of the moon. The moon revolves about the earth in an elliptical path, causing the distance separating the two bodies to vary by as much as 42,200 km (26,200 mi). This difference results in a slight but noticeable change in the apparent size of the moon in the sky, from 33' 31" in angular diameter when nearest the earth (perigee) to 29' 22" when farthest from the earth (apogee).

Apparent size of the sun. The earth's path about the sun is also elliptical, with the distance separating the two bodies varying by nearly 5.1 million km (3.2 million mi). This results in a change in the apparent size of the sun in the sky, from 32' 36" of angular diameter in early January, when the earth is closest to the sun (perihelion), to 31' 31" when the earth is farthest from the sun (aphelion) six months later.

Total solar eclipse. For a total eclipse of the sun to occur, the moon must pass directly in front of the sun from an observer's vantage point on the earth. In addition, the apparent size of the moon must be equal to or greater than the apparent size of the sun to allow the moon's visible disk to completely obscure the sun from view.

For the hypothetical observer on the moon, the small black disk of the moon's shadow would be seen superimposed on the full earth and racing across it from sunrise to sunset points in a generally west to east direction but usually with a significant northward or southward component. The moon's umbral shadow is much smaller than the earth's; at its largest it has a width of only 270 km (168 mi) at the earth's surface.

Annular solar eclipse. When the moon passes directly in front of the sun, but the apparent size of the moon is less than that of the sun, an annular eclipse of the sun takes place. The moon appears as a hazy gray disk surrounded by a ring or "annulus" of light from that part of the sun which the smaller lunar disk has left uncovered.

Another way of thinking about an annular eclipse is to envision the moon being too far from the earth for its umbral shadow, which varies between 367,400 km (228,300 mi) and 379,900 km (236,100 mi) in length, to reach the earth's surface. In fact, it can fall short of reaching the earth's surface by more than 32,000 km (20,000 mi).

The moon's penumbral or negative shadow is at most 370 km (230 mi) wide at the earth's surface, only slightly larger than its umbral shadow and considerably fainter.

Annular–total solar eclipse. The earth is not a true sphere, but has bulges that give it more the shape of a pear. When the apparent sizes of the sun and moon are very nearly identical, these bulges allow the moon's umbral shadow to touch the earth's surface briefly along the path of an otherwise annular eclipse, resulting in an annular eclipse for observers at most locations and a total eclipse at others. Extremes of terrain, especially where the path crosses mountains, as well as changes in the moon's distance from the earth during the eclipse, can be minor contributing factors as well, but more frequently an aircraft

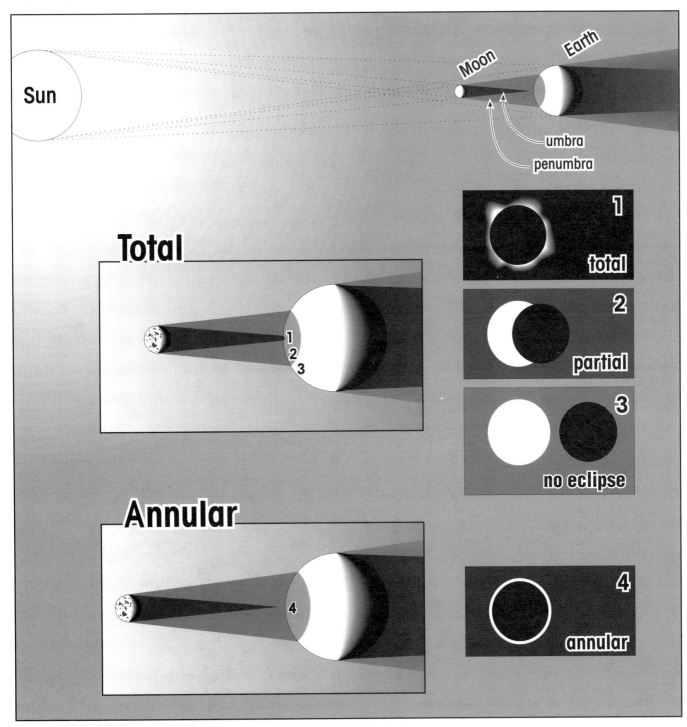

Figure 2-3 Solar Eclipses. The moon's shadow in relation to annular, partial and total solar eclipses. (Due to space limitations, the sun is shown proportionally closer to the earth than it is in reality, therefore the shadow sizes and shapes are not to scale.) *Illustration by David Frantz.*

is required to carry observers high enough to encounter the umbral shadow and experience totality while most of their ground-based counterparts along the eclipse track are experiencing an annular eclipse.

From a lunar vantage point, noting any fluctuations in the darkness of the penumbral/umbral shadows as they cross the earth's uneven terrain and cloudtops might prove interesting.

Partial solar eclipse. When the moon crosses the sun's disk but is too far north or south of the sun's center to completely obscure its light, a partial eclipse of the sun occurs.

This least-popular type of solar eclipse for earth-bound observers would offer even less to their future lunar-based counterparts as the moon's umbra misses the earth entirely.

Orbital motions of earth and moon. As the earth revolves about the sun once every 365.26 days, an earth-based observer would see the sun slowly drift from west to east among the background stars. The path of this drift, through the twelve constellations of the classical *zodiac*, is called the *ecliptic*.

The moon orbits the earth once every 27.32 days, a period referred to as a *sidereal month*. During this period, the moon returns to the same point, with respect to the earth and stars, that it started from.

During this period, however, the earth has been revolving about the sun and carrying the moon along with it. When the moon returns to its original position with respect to the earth, it is not quite in its original position with respect to the sun. The sun's position has shifted some 27 degrees with respect to the moon. For the moon to return to the same phase it had one sidereal month earlier, it must make up those 27 degrees by traveling an additional 2.209 days.

This longer period, 29.53 days, is called the *synodic month*. Since it determines the recurrence of any given lunar phase, it also determines when eclipses may occur. Since a lunar eclipse must occur near the time of full moon and a solar eclipse at new moon, it would seem logical to expect an eclipse of each kind every synodic month. However, due to the tilt of the plane of the moon's orbit, such is not the case.

The moon's orbital plane. The moon's orbit about the earth is tilted (inclined) by about 5 degrees with respect to the earth's orbit about the sun. As a result, the moon crosses the sun's path through the heavens, the ecliptic, twice each synodic month. These two points of crossing are called *nodes*.

The moon crosses the ecliptic moving northward at the *ascending node* and southward at the *descending node*. An eclipse is only possible when one end of a line connecting the ascending and descending nodes of the moon's orbit, called the *line of nodes*, points toward the sun.

An umbral eclipse of the moon may take place as much as 12 days on either side of a node passage (which is to say, 12 degrees on either side of a node, since it takes just about one day for the earth to travel one degree in its path). This total of 24 available days is less than one synodic month, however, there is at most only one opportunity per node passage for a total or partial umbral lunar eclipse to occur when the full moon occurs during the 24 day period. There are at most three opportunities in a calendar year for the line of nodes to point at the sun (see *Eclipse seasons*, next section). This means there are at most three possible total and partial lunar eclipses which can occur in a calendar year. Sometimes no lunar eclipse occur in a calendar year. So lunar eclipses are considered somewhat rare.

It is possible for two penumbral lunar eclipses to occur at solar-aligned node passages, so if a total eclipse occurs at the third passage and two penumbral eclipses occur at each previous node passage, the maximum number is five lunar eclipses in a calendar year. However, penumbral eclipses are generally considered to be of little or no scientific value and are frequently excluded from references listing lunar eclipses.

For a total solar eclipse to occur, the sun must be close to one of the two nodes of the moon's orbit at the same time that the moon is crossing that node. Solar eclipses may range up to 18 days on either side of a node passage (or 18 degrees on either side of a node). Since the total of 36 available days is greater than one synodic month, it is possible for two solar eclipses to occur at each node passage if two new moons occur during that time. Because of this spread, a solar eclipse of some type must occur during each of the node passages since at least one new moon occurs during the 36 days.

At times other than nodal encounters, the moon would appear to pass north or south of the sun (or of the earth's shadow) and no eclipse would occur. This important qualification is what gives rise to *eclipse seasons*, times of year when eclipses can occur.

Eclipse seasons. When the line of nodes of the moon's orbit approaches right angles to the sun-earth line, the moon is crossing a node while in the first or third quarter phase. At full or new moon the moon is passing north or south of the sun and no eclipse is possible. Therefore, solar and lunar eclipses can only occur at intervals roughly six months apart when the line of nodes is aligned with the sun and the moon is near one of its nodes. These positions are referred to as eclipse seasons.

However, the line of nodes itself is slowly rotating clockwise (westward) about the ecliptic at a rate of some 19 degrees per year, carrying the ascending and descending node positions with it. This so-called *regression of the nodes* causes the inclination of the moon's orbit relative to the earth's equator to vary between 18.5 degrees and 28.5 degrees (the earth's axial tilt of 23.5 degrees ± 5 degrees). The time for one complete circuit of the ecliptic is about 18.6 years. This phenomenon is caused by the gradual but continuous change in orientation of the moon's orbit as a result of gravitational effects of the sun and, to a lesser degree, the earth and other bodies in the solar system, and is responsible for the *Metonic* and *saros cycles* which allowed for the earliest eclipse predictions.

Because of this regression of the nodes, a 19-degree clockwise shift in the position of the line of nodes occurs during the calendar year. Since the earth moves about one degree per day around the sun, the time it takes for the same node to make its second consecutive alignment with the sun, or the *eclipse year*, is shorter than the calendar year by just over 19 days (more precisely, the eclipse year is 346.62 days in length or 341.64 degrees of ecliptic travel). Therefore, if an eclipse season starts out in January, it is possible for that calendar year to experience three eclipse seasons, the last one falling in December.

Eclipse seasons occur just over 19 days earlier each year. In 1994, eclipse seasons fell in May and November; by 1997 the seasons fall in March and September. Every three years eclipse seasons occur about two months earlier. The last year having three eclipse seasons was 1992 (January, June and December); three eclipse seasons occur in the same calendar year in 2000 (January–February, July and December), and 2009 (January, July and December).

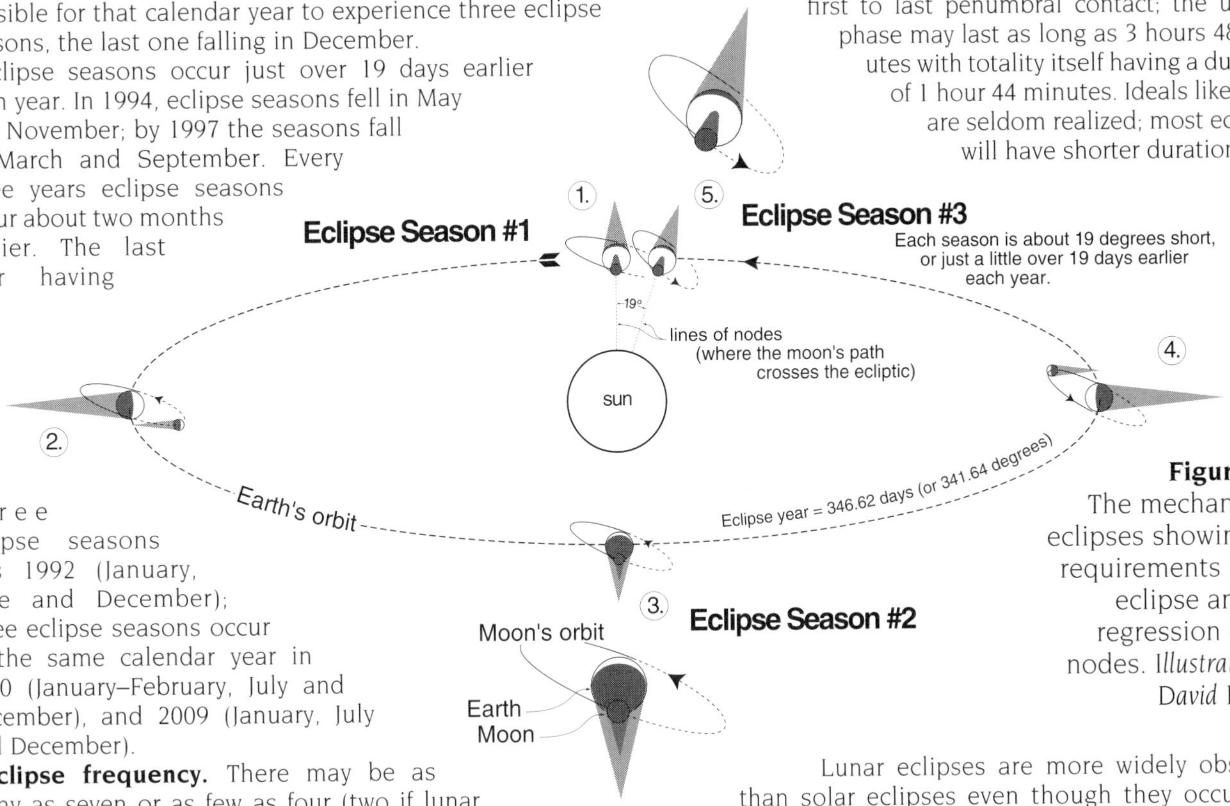

Eclipse frequency. There may be as many as seven or as few as four (two if lunar penumbrals are disregarded) eclipses in any given year. In a year having the maximum number, they may occur in any of the four combinations which follow:
1. Five lunar (at least four of which must be penumbral) and two solar (both central);
2. Two lunar (both total) and five solar (at least four of which must be partial);
3. Four lunar (at least two of which must be penumbral) and three solar (at least one of which must be central);
4. Three lunar (at least one of which must be total) and four solar (at least two of which must be partial).

There were seven eclipses in a single year in 1917 (Combination 4 above), 1935 (Combination 2), 1973 (Combination 3), and 1982 (Combination 4, with three total lunar and four partial solar eclipses visible). Seven eclipses will also occur in the same year in 2038 (Combination 3, with four penumbral lunar, two annular solar and one total solar eclipses visible), 2094 (Combination 4), 2103 (Combination 3), 2132 (Combination 1), 2159 (Combination 4), and 2206 (Combination 2).

It is interesting to note that the last time there were three total lunar eclipses in one calendar year was in 1917 (on 8 January, 4 July and 28 December); this will not occur again until 2485 (on 1 January, 28 June, and 21 December).

The maximum possible number of solar eclipses in a calendar year, five, last occurred in 1935 (four partials and one annular); it will next occur, in the same combination, in the year 2206.

Two solar eclipses have occurred in the same calendar month in December of 1880; this will recur on 1 July and 31 July in the year 2000, but not again until 2206.

Visibility and duration of eclipses. An ideal central total lunar eclipse may last 5 hours 44 minutes from first to last penumbral contact; the umbral phase may last as long as 3 hours 48 minutes with totality itself having a duration of 1 hour 44 minutes. Ideals like these are seldom realized; most eclipses will have shorter durations.

Figure 2-4 The mechanics of eclipses showing the requirements for an eclipse and the regression of the nodes. *Illustration by David Frantz.*

Lunar eclipses are more widely observed than solar eclipses even though they occur less often, because the people on the entire hemisphere of the earth where night has fallen and the moon is above the horizon can observe the full moon slip through the earth's shadow at the same time.

Solar eclipses are much more localized than lunar eclipses, and have been observed by far fewer people. The moon's shadow falling on the earth is much smaller than the earth's shadow projected out into space, and one must be directly in the path of the moon's shadow to experience the total solar eclipse. Partial solar eclipses are more common from any given site, and have been much more widely observed.

A solar eclipse may last up to four hours, but totality may theoretically not exceed 7 minutes 58 seconds (the 25 June 2150 eclipse will last 7 minutes 14 seconds, longest in the 13 centuries preceding it). Modern supersonic aircraft can extend totality for observers to over an hour by matching the speed of the moon's shadow. Annularity is limited to 12 minutes 24 seconds for a ground-based observer. Average durations, as with lunar eclipses, are usually considerably less than these extremes.

Magnitude of eclipses. The fraction of the sun's diameter obscured by the moon at greatest phase, or the fraction of the moon's diameter obscured by the earth's penumbra or umbra at greatest phase defines an eclipse's *magnitude*. Magnitudes greater than or equal to 1.000 identify total obscuration.

The saros. The ancient Egyptians and Chaldeans recognized that after a period of just over 18 years, the alignment of nodes and the relative positions of the sun and moon repeat themselves. Therefore, solar eclipses separated by this interval will have similar durations of totality and other characteristics as a previous eclipse.

This saros, a word which meant "repetition" to the ancient Babylonians, occurs whenever the 29.53–day synodic month and the 346.62–day eclipse year intervals fall on the same day. An interval of 223 synodic months, equal to 6585.32 days (18 years 10.64 days), provides for such a saros interval since it nearly equals 19 eclipse years (6585.78 days).

An eclipse repeating itself after the passage of one saros (223 synodic months) will occur 0.32 day later. Since the earth rotates 360 degrees in one day, this translates to a 115.2–degree westward shift in longitude and a starting time that is 0.32 day (7.7 hours) later for this next eclipse in any given saros. Even after three successive saros cycles, an eclipse has shifted only 0.96 day and is still 14.4 degrees east of its original starting point.

Each saros bears a number, which is shared by all eclipses belonging to that cycle, following a system proposed by the Dutch astronomer G. van den Bergh in 1955. For example, the total solar eclipse of 3 November 1994 belongs to saros 133 and is the 44th of 72 eclipses belonging to that particular series. The previous solar eclipse series member was on 23 October 1976, while the next one occurs on 13 November 2012.

A saros series may be completed in as few as 1200 or as many as 1500 years depending upon the geometric circumstances of the first eclipse in that series. The average saros series has 73 eclipses and lasts 1315 years.

The inex series. This repetitive series lasts 23,000 years and includes 780 eclipses visible from alternately opposing latitudes. Each eclipse in an *inex* series recurs at intervals of 358 synodic months. A subsequent eclipse in an inex series occurs only 4.3 minutes earlier (a 0.0411–degree eastward shift), so it appears at nearly the identical longitude but in the opposite hemisphere as the previous one. It may be said that each eclipse belongs to a saros and each saros to an inex.

The Metonic cycle. The Greek astronomer Meton, who lived around 430 B.C., discovered that a cycle of 235 *lunations* (6939.69 days) coincided very closely with an interval of 19 *solar (tropical) years* (6939.60 days).

The significance of Meton's discovery is that the moon undergoes the same series of phase changes on the same calendar date in intervals of 19 years. Therefore, a full moon occurring on 19 October 1994 would be repeated on 19 October 2013.

Four Metonic cycles are sufficient to make a difference of a full day in this recurrence of lunar phases, however, so the cycle is not a very efficient long–range predictor for recurring phases.

Meton's cycle proves even less efficient as an eclipse predictor. In 1994, a penumbral eclipse accompanies the 18 November full moon; there was a total eclipse at the full moon on that date in 1975, partial eclipses in 1956 and 1937, but no eclipse in 1918. There will also be no lunar eclipse in November of 2013 following the passage of the next Metonic cycle.

References

Bishop, Roy L., ed. **Observer's Handbook 1994**. Toronto: The Royal Astronomical Society of Canada, 1993.

Couderc, Paul. **Eclipses And Their Sequences: Leaflet No. 384–June, 1961**. San Francisco: Astronomical Society of the Pacific, 1961.

Espenak, Fred. **Fifty Year Canon of Lunar Eclipses**. Cambridge: Sky Publishing Corporation, 1989.

Espenak, Fred. **Fifty Year Canon of Solar Eclipses**. Cambridge: Sky Publishing Corporation, 1987.

Liu, Bao–Lin and Fiala, Alan D. **Canon of Lunar Eclipses 1500 B.C.–A.D. 3000**. Richmond, VA: Willmann-Bell, Inc., 1992.

Mitchell, S. A. **Eclipses of the Sun**. New York: Columbia University Press, 1923.

Smith, Michael R. **TotalEclipse™** (IBM–compatible software). Pittsburgh: Zephyr Services, 1993.

Zirker, J. B. **Total Eclipses of the Sun**. New York: Van Nostrand Reinhold Company, 1984.

Eclipse Observing and Vision Safety

Solar observing. The sun emits intense radiation in the infrared, visible and ultraviolet bands of the electromagnetic spectrum. We protect ourselves, at least partially, against the infrared and ultraviolet wavelengths of light with hats, sunscreen lotions, or by seeking shelter out of the sun's direct rays. We deal with the visible wavelengths of light, reflected from bright surfaces, with sunglasses. Sunglasses, however, do not provide protection to the eyes from looking directly into the sun.

Our eyes are very sensitive to infrared, ultraviolet and intense visual solar radiation, and can easily sustain temporary or permanent damage—even blindness—from staring at the sun. The condition is called solar retinopathy (i.e., burns on the retina) and symptoms may not appear for hours after staring at the sun. In mild cases, symptoms may disappear; in moderate or severe cases, permanent loss of vision or portions thereof result.

This condition is not unique to a solar eclipse, contrary to what many are led to believe by well–meaning astronomers and medical professionals, but is caused by looking at the sun. The emphasis on safety for the impending solar eclipse results in unfortunate wording in media interviews that so convince people not to look at the sun "during the eclipse" that they mistakenly believe it is safe to do so once the eclipse is over.

The sun at totality. During—and only during—totality, it is perfectly safe to look at the totally eclipsed sun, without any filtering devices, even through telescopes. Obviously, during totality it is the new moon which is being observed in silhouette against the sun's outer atmosphere, and it is safe to view the moon without special filters.

However, great care must be taken to insure that such viewing does not begin until the last bead of direct sunlight winks out, and even greater care must be taken to insure that observations end before the edge of the sun reappears from behind the moon's limb at the end of totality. This latter task is best left to a reliable volunteer timekeeper, or tape–recorded countdown, to warn of the approach of third contact at the observer's site, than to the preoccupied observer's personal judgment.

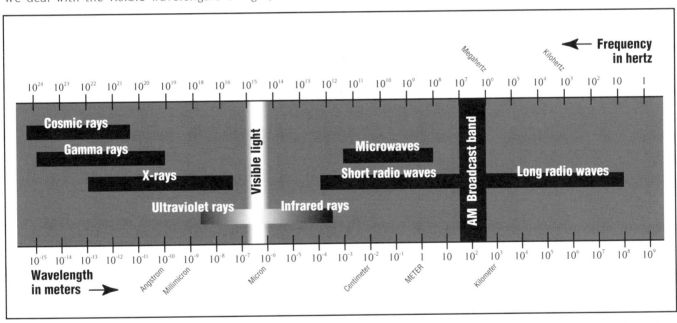

Figure 3-1 The Electromagnetic Spectrum showing the location of visible light with ultraviolet and infrared rays on opposite wavelengths to the visible spectrum. *Illustration by David Frantz.*

Lunar observing. The moon, whether in or out of eclipse, presents no danger to the eyes. Thoughtful amateur astronomers should not assume that lay persons or the media know this, and should be prepared to field their concerns and inquiries at times of lunar eclipses with assurances and explanations rather than ridicule. The moon, which is full at times of lunar eclipse, can be uncomfortably bright during the early and late stages of an eclipse, however, and filters can enhance the observing experience.

What's safe and what isn't? *It is never safe to stare directly at the sun with the unaided eye.* At times this seems to be unavoidable. Driving to and from work on east–west highways can expose commuters to the rising and setting sun for extended periods of time. Tinted windshields and sunglasses may handle the casual glances, but are poor substitutes for avoidance. Effective use of sun visors or similar blocking devices is best.

Many believe that viewing the sun's reflection off a body of water or piece of dark glass or metal automobile hood is safe; it is not, for the sun's reflected light can also cause eye damage.

The safest way to observe the sun with the unaided eye is to project an image of the sun onto another surface; this **projection method** guarantees that at no time is the eye exposed directly to solar radiation.

The next–safest way is to obtain approved solar **viewing filters** from a reputable science supply house or astronomical company. Inspect them carefully for flaws before each use, then discard and replace them after a reasonable period of time.

It is never safe to observe the sun through any kind of optical instrument without suitably approved astronomical filtering devices securely attached. This includes binoculars, small spotter scopes, and camera viewfinders and lenses.

Projection method. The use of projection to view the sun's image indirectly is by far the safest approach to solar and solar eclipse observing. This method may be applied either with or without optical aid, although the former presents a much larger image and more rewarding appearance. Both approaches will be discussed, along with their advantages and disadvantages.

Pinhole projectors. A simple pinhole projector can be made from a cardboard shoebox. Cut a small opening about 2.5 cm (1 in) square in the center of one end of the shoebox, and tape a piece of aluminum foil over the opening. Make a small opening in the center of the aluminum foil with a pin. The sun's image will be projected onto the opposite end of the shoebox, where it can be viewed by an observer.

Viewing may be enhanced by leaving the lid on the shoebox and cutting a small viewing slit in the side of the box near the end where the sun's image is being projected, and placing the eye near this slit. The sun's image may then be seen against a relatively dark background. The best view is obtained with the observer's back to the sun and the aluminum pinhole held over the shoulder, guaranteeing that the observer is not tempted to glance at the sun while viewing the eclipse.

A similar device can be made by replacing the cardboard shoebox with a large mailing tube. The pinhole can be made in an aluminum foil cap taped to one end. A paper viewing screen, made translucent by a drop or two of cooking oil, may be taped to the other end in lieu of cutting a viewing slit in the side of the tube.

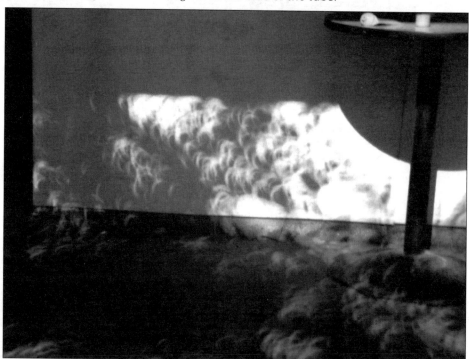

Photograph 3-2 Nature provides an excellent pinhole projector: the overlapping leaves of a tree! 10 May 1994 annular, Baja California, Mexico. *Photograph taken by Mike Reynolds.*

Mirror projectors. It is possible to view the sun from indoors with a simple mirror device. Select a window facing the sun. Tape a piece of paper, with a 2.5 cm (1 in) hole cut in its center, onto the window, then cover the rest of the window with dark cloth or sheets of newspaper.

Next, place a mirror against an opposite wall so that the sun's light coming through the opening in the window hits it. Reflect the sun's light from the mirror onto a piece of paper or white cardboard attached to the wall beneath the window.

Finally, cut a 2.5 cm (1 in) hole in another piece of stiff cardboard, tape a piece of aluminum foil over the hole, and make a pinhole in the center of the aluminum foil. When this pinhole card is moved between the mirror and the cardboard beneath the window, an image of the sun should come into focus on the wall beneath the window.

The mirror will have to be moved to follow the sun's motion across the sky as the eclipse progresses, but this method assures comfortable indoor viewing of the partial phases as long as the sun is shining through the window opening.

A similar method, recommended by the Royal Astronomical Society of Canada in their *Observer's Handbook 1994*, is to cover a small pocket mirror, except for a small opening about 6 mm (1/4 in) square, and position it where the sun's rays can reflect off the opening and into a darkened room. The reflected spot will be a pinhole image of the sun's disk. Try varying the size of the opening, and the distance you project the image, to maximize image size, sharpness and brightness.

Optical projectors. A pair of binoculars or a small telescope can be used without filters to project an image of the sun onto a screen, a hand–held square of cardboard or posterboard. The instrument being used should be securely mounted and carefully monitored at all times to make certain that curious onlookers do not attempt to look directly through the instrument and damage their vision. For public observing sessions, amateur astronomers are

Photograph 3-3 A number of options can be considered for optical projection. Sun Spotter II, a commercially–available product, produced excellent images of the 10 May 1994 annular from Baja California, Mexico. *Photograph taken by Mike Reynolds.*

urged to rope off the instrument and the screen to keep spectators from placing their eyes between the two.

A round cereal box, such as an oatmeal container, can also be used. Simply slip one end of the box over the eyepiece end of a telescope and view the projected image at the other end through a viewing flap or hatchway cut in the side near the end opposite the telescope.

Viewing filters. Whenever an observer intends to use filters to view the sun or a solar eclipse directly, there are inherent risks in the procedure even when the filters themselves are capable of blocking all of the sun's harmful rays and dimming the visible light to a comfortable level.

Filters which are used in conjunction with telescopes or other optical devices present the greatest risk to observers, for the sun's light is being intensified and magnified by the optics of the instrument. Anyone who has ever focused the sun's light through a magnifying glass and ignited a leaf or a piece of paper has personal knowledge of the sun's ability to damage the eyes.

The two approaches to filtering the sun's rays for direct viewing are the rear–mounted and front–mounted methods, so named for their placement with respect to the optical path of an instrument.

Rear-mounted filters. The most dangerous type of filter is one that is placed at the eyepiece end of binoculars or telescopes, for these are taking the full force of the magnified image of the sun.

Even though they are capable of safely filtering out the sun's harmful rays, their rear–mounted placement subjects them to intense heat from the sun.

Photograph 3-4 The best thing to do with rear–mounted eyepiece solar filters: *throw them away!*

As they heat up, their glass expands within their mounting cells. If they have internal flaws or if their mounting cells are too tight to allow for expansion as they heat up, they will crack; many will crack even if well made and properly mounted. An observer looking through one when it cracks is unlikely to react quickly enough to withdraw the eye before sustaining serious injury. Unfortunately, they tend to be the most common type among beginning amateur astronomers, since such filters frequently come with the popular and inexpensive imported 50–mm to 60–mm (2–in to 2.4–in) refracting or 76 mm (3–in) reflecting telescopes distributed through popular department–store chains. The best advice is to throw them out and obtain a safer, front–mounted filter.

Front–mounted filters. These filters are placed over the front of binoculars, telescopes and camera lenses and filter the sun's light before it ever enters the optical system. Therefore, the heat stays away from the instrument and the observer's eyes. For small instruments, they can be full–aperture filters, meaning they have the same surface area as the telescope lens or mirror; for bigger instruments it is usually desirable to mount a smaller filter into a cell which fits over the end of the telescope, effectively reducing the aperture and saving money on the filter as well.

These filters may be made from metal-coated glass, mylar or plastic material, and provide adequate safety and pleasing views. However, they are not without safety concerns which should be constantly addressed.

The coatings on glass filters can deteriorate with time or become scratched; the best of these have the coated surfaces sandwiched between two pieces of glass, protecting them from the elements and from damage due to handling. The mylar and plastic filters can also suffer surface degradation, but are more prone to pin pricks due to handling in use. Some mylar and plastic filters also sandwich their coated surfaces for added protection and longevity, but these types are generally inexpensive enough to warrant replacement after a reasonable amount of use.

Photograph 3-5a, b Front-mounted filters. Above, a Thousand Oaks coated glass filter. Below, a selection of Tuthill mylar filters.

Front-mounted filters present an additional risk to the observer intent on direct solar viewing. They are only as safe as the method used to securely mount them to their instrument. Observers who rely on masking tape are risking their vision at every observing session. Even filter mounts which seem to fit snugly over the optical tube have been known to fall or be knocked off.

The best approach, and one that is absolutely essential around the general public or playful school students, is to fashion a mount that attaches snugly and is secured in a way that only you can quickly release at totality. The finder scope should also be equipped with a securely-mounted filter or be removed from the telescope entirely.

Distributors of safe front-mounted filters are listed in Appendix 5, but sources and addresses frequently change; potential purchasers should consult periodicals such as **Sky & Telescope** and **Astronomy** magazines for up-to-date listings and prices. They are advised to plan such purchases at least several months prior to an eclipse to guarantee delivery in ample time to practice observations and photography on the uneclipsed sun.

Other approaches. The classic system of telescopic solar observing employed a combination of an unsilvered glass secondary or diagonal mirror, called a Herschel Wedge, which ejected all but a tiny fraction of the sun's light from the telescope, and a rear-mounted sun filter of Number 4 density dark glass mounted between the eye and eyepiece. While still a reliable approach, the availability of inexpensive full-aperture front-mounted filters makes changing secondaries and mounting sun filters unnecessarily time-consuming unless you have a telescope dedicated solely to solar observing.

For the naked-eye, the use of a Number 14 welder's glass provides adequate protection from infrared as well as visible light to allow for safe viewing. They are not recommended for use with telescopes or binoculars.

The Eastman Kodak Company quotes medical authorities as recommending neutral-density filters of metallic silver and having at least a 6.0 density for naked-eye use. To make these, unroll a newly-opened roll of black-and-white panchromatic photographic film containing *silver*, such as Kodak Plus-X or Tri-X, to direct sunlight. Then roll it back onto its spool or into its cassette and have it developed for *maximum density* according to the manufacturer's recommendations. When it has been processed, cut it into equal lengths long enough to cover one or both eyes, and tape *two thicknesses* of the film together for viewing. Mounting the film in a cardboard frame makes a convenient holder and protects against fingerprints. The National Society to Prevent Blindness correctly points out that non-professionals frequently misunderstand and misinterpret these critical instructions for making photographic film filters. *Color films and newer black-and-white films which do not contain silver are not safe to use, nor are undeveloped film or developed negatives with photographic images on them. Do not use this approach unless you know what you are doing!*

Unfortunately, film filters are generally too dark to use as a photographic filter, and less-dense photographic filters do not provide adequate protection to the eyes for even the briefest of glances at the sun through a camera viewfinder. Also, photo processors rarely develop black–and–white film in house anymore, necessitating use of more expensive labs, lengthy shipping delays, or, if you are so equipped, developing the film yourself. Again, the best approach is to use the front–mounted professional sun filters instead, which are safe for visual use and yield pleasing photographs. Planetaria and museums often sell filters especially made for eclipse viewing.

Unsafe methods reemphasized. Never view the sun, either in or out of eclipse, with the unprotected eye. The only exception is totality, when none of the sun's bright surface is visible. Sunglasses, crossed polarizing filters, color negatives, color transparencies, color films, black–and–white film containing no silver, undeveloped film, bottles of colored or dyed water, reflections of the sun in dark glass or standing water, and smoked glass do not provide adequate protection and are unsafe solar viewing options. Serious eye damage can accompany the use of any of these methods. Shun all rear–mounted filters on optical instruments. Finally, even approved–safe direct–viewing filters should be inspected carefully for flaws, scratches and damage before risking eye injury by their use.

Photograph 3-6 The importance of safe solar viewing at all times, whether during a solar eclipse or solar observing in general, cannot be overemphasized. Mike Martinez (left), and Jeremy Reynolds demonstrate the use of naked–eye mylar and film filters at the 10 May 1994 annular eclipse.

Peanuts reprinted by permission of UFS, Inc.

References

Bishop, Roy L., ed. **Observer's Handbook 1994**. Toronto: The Royal Astronomical Society of Canada, 1993.

Brewer, Bryan. **Eclipse, 2nd ed**. Seattle: Earth View, Inc., 1991.

Brown, Sam. **All About Telescopes**. Barrington, NJ: Edmund Scientific Company, 1972.

Eastman Kodak Company. **Astrophotography Basics**. Rochester, NY: Eastman Kodak Co., 1988.

Eastman Kodak Company. **Solar Eclipse Photography For The Amateur, 2nd ed**. Rochester, NY: Eastman Kodak Co., 1963.

Moeschl, Richard. **Exploring the Sky: 100 Projects for Beginning Astronomers**. Chicago: Chicago Review Press, 1989.

National Society to Prevent Blindness. **Solar Eclipse Safety Information Sheet for Science and Astronomy Teachers, Astronomy Clubs, Planetariums, Science Centers and Eye Care Professionals**. Schaumburg, IL: National Society to Prevent Blindness, 1994.

Expedition Planning

Introduction. A successful eclipse expedition depends upon early and careful planning. You may choose to organize your own independent expedition for your immediate family and friends, a larger offering for your club, or you may prefer to organize or participate in a major commercial venture. This chapter discusses these options and offers a planning approach which has proven successful for others.

Commercial ventures. In the last four decades of the twentieth century, the availability of economical land, sea and air transportation has opened eclipse sites in even the most far-flung corners of the earth to amateurs and sightseers alike. Many travel agencies, clubs and enterprising individuals organize tours to take advantage of every eclipse opportunity. Even eclipses which take place entirely over the ocean have yielded to cruises aboard luxury liners which are guided to the region of best weather by artificial satellite imagery. Both include side tours at "ports-of-call" having attractions of general interest to tourists.

Many commercial ventures, especially shipboard cruises, include seminars on astronomy, photography, navigation, meteorology, oceanography, and other disciplines. Heading up these seminars are astronauts, science-fact and science-fiction authors, magazine editors, and leading professionals in their fields. Although such expeditions carry higher costs, the ready availability of famous personalities to chat and dine informally with you is an appealing advantage.

Educational opportunities are plentiful as well. For an additional tuition fee, many expeditions offer the option of receiving undergraduate or graduate college credit for participating in an eclipse seminar and completing a project or paper. Other options frequently include continuing education credit and teacher inservice credit for your participation.

Most major organizers of popular eclipse tours enjoy phenomenal success at getting their participants to a clear site for the eclipse, while handling all essential logistics smoothly and with a minimum of bother for their clients. Two words of advice to those considering launching their own commercial ventures, however; involve an experienced travel agency and obtain adequate liability insurance to protect your participants and yourself from potential accidents and financial crises.

Advertisements for commercial eclipse ventures begin appearing in the pages of **Astronomy** and **Sky & Telescope** magazines as early as 18 months in advance of an eclipse. You might also contact a local travel agent, astronomy club, planetarium, observatory, college science department, or the Astronomical League. Make reservations early; the better tours fill up quickly.

Independent expeditions. Many individuals choose to organize their own expeditions. Others, affiliated with a club or school, will band together in small groups to tailor-make an expedition to their own interests.

With a little help from a travel agent or the travel desk of a major airline, substantial savings may be realized over the costs of commercial tours and they can be every bit as exciting and successful as professionally organized tours. There are many details which must be dealt with to avoid later headaches, however. These are stressed in the next section.

ECLIPSE planning method. One planning approach with a proven track record for solar eclipses, the type you are most likely to travel long distances to see, we call, appropriately, the **ECLIPSE** method for the first letters of the words **E**stablish, **C**hronicle, **L**ocalize, **I**nstrument, **P**ractice, **S**ustain and **E**valuate. It is presented here as a guide for those embarking on their first expedition.

Establish: It is important to establish clear goals, which are within your skill, equipment and finances to achieve, from the very beginning. It is wise to start planning a year or more ahead of the eclipse you want to see. In this stage, you have several decisions to make:

1. Where will you go to see the eclipse? This can be as near as the front yard or as far as halfway around the earth. If a long, expensive trip is required, it is important to gather weather and climate records for the areas you are considering (see **Weather: The Major Eclipse Unknown**). **Sky & Telescope** and **Astronomy** magazines usually publish weather information and prospects for sites along an eclipse path more than a year before the event to help you select the best site. If you plan to conduct serious scientific work, you will also need topographic maps to pinpoint your exact latitude, longitude and elevation once you are at your site. Most countries have them, but you will need to order them well in advance. If you're flying, book round-trip air reservations early and don't plan to arrive on eclipse day; flight delays or misdirected luggage can doom you! Special-fare

Weather: The Major Eclipse Unknown

Good information on the climatology of potential eclipse viewing sites is readily available, and site decisions should rely on these data. Of course, if you really want to see Thailand instead of India, you may choose to reject climatological recommendations, but at least you will be aware of the risks.

A small mobile group should choose alternative routes along the eclipse path and plan for moves of several hundred kilometers in middle and high latitudes (50 to 100 kilometers for tropical locations) whenever borders, coastlines and available roads permit. Long moves may take you away from your hotel for a night or force you to do without a regular lunch or breakfast stop.

Deciding whether to use the main site or one of the escape routes requires good weather advice. CNN is available globally, but the forecasts are usually rudimentary, for 24 hours only, and concentrate on major cities. Watch for an hour or two in order to see enough of the satellite image to make a decision.

Most television satellite images are heavily processed and show only the main cloud systems. There may be quite extensive cloud patches which are completely invisible. Try several channels, even if you don't understand the language. Watch with several people, each concentrating on a different part of the image; a group decision is usually more accurate and, if not, at least you can all share the blame!

Try to contact the local airport weather offices, using hotel staff if need be and visit them if possible. Question them carefully and at length and carry along a few extra eclipse filters, or a t–shirt to hand out. Don't rely on unspecialized advice from locals; they know general trends well enough, but don't think the afternoon thundershowers really detract from an "always sunny" description all that much. Even weather offices care little about cirrus level cloud since it has no effect on the airport.

If you can't get advice from anyone, keep an eye on the sky yourself. A convective cloud will dissipate, as long as it hasn't grown to the size where it's beginning to rain before the eclipse. Watch the motion of clouds very carefully. In the tropics wind speeds are usually very light, and cloud motion difficult to see, especially if the cloud is forming and dissipating along the way. Use a telescope to watch the cloud edges; it will magnify the motion dramatically. Fog can form on slopes from a combination of uphill motion and eclipse cooling, as I learned during the Indonesia eclipse of 11 June 1983. A third of our group headed up the volcano road seeking drier skies at higher altitude. While the clouds disappeared for those of us below, the higher chasers had to run for a few holes in the gray clouds which spread up the volcano slopes as the air cooled. Uphill winds should be religiously avoided at eclipses, especially in the tropics. If possible, sites should be chosen on the downward side of the terrain, though I tend to recommend against hilly or mountainous sites as a matter of principle. It's too easy for cloud to form there, and it's too difficult to get out of the way.

A British group watching the 11 July 1991 eclipse from the Pacific side of Baja made the mistake of staying in an area where they had heavy fog in the early morning hours. Even though the eclipse was at noon and the fog had long gone, fog forms when a moist surface layer cools, exactly the effect of an eclipse. The fog rolled in again, cutting a seven minute eclipse to one minute. Other groups, uphill and away from the ocean, saw the complete eclipse.

My next eclipse I intend to take along a small receiver for cloud images from U.S. and Russian polar orbiting satellites. They are a bit pricey, but can be used anywhere in the world. Capture several images then watch the cloud motion. Examine both the visual and infrared images; clouds look much more extensive in the infrared. Plot the patches on a background map and keep track of their speed and direction. Extrapolate cloud edges and holes into the future, and pick your site accordingly. Of course, a good meteorological background is helpful in understanding the imagery. Global satellite images are available on the Internet. If you can get access, you can download the images and view them yourself.

Finally, don't overlook advice from home. I briefed expeditions in Uruguay for the 30 June 1992 sunrise eclipse from my home weather office in Winnipeg. Almost the entire track was clear except for the area closest to the center line (which didn't quite touch land). I passed that information to an expedition in the area, and advised them to move back inland about 50 kilometers. Darkness, lack of roads, and no acceptable observing site caused them to miss the eclipse. To the best of my knowledge, no one saw that eclipse, but the clear skies were there for the taking if a few seconds of totality had been sacrificed. I suspect that I could brief an expedition anywhere in the world now, just sitting at my home. Weather is still the bug–bear of eclipses, but modern communications is making it more tractable.

Jay Anderson, Meteorologist
Environment Canada

reservations made well in advance are the least expensive, but generally carry cash penalties for changes and most are non–refundable. Examine all your options.

2. What will you need to get there? Passports, visas, immunization records, and adequate health, accident and liability insurance can take time to obtain; make certain that you and every member of your team has arranged for them early. Customs can present problems on international trips; have several copies of a detailed list of all equipment, including camera lenses and bodies, along with their serial numbers, for filling out or attaching to customs forms. Photographs of the equipment you're carrying can be useful in the event of theft or loss. Familiarize yourself with the region you'll be visiting by examining tour books and travel brochures; your travel agent can help here. Careful research will suggest other, more utilitarian, needs such as special clothing and footwear, electrical adaptors, appropriate film, tapes and batteries, and personal dietary or medicinal needs. If you're going with an organized group or travel agency, many of these details will be coordinated for you; if you're on your own, don't overlook them.

3. Where will you stay? Camping out may be an option; if so, make reservations or arrangements early. If not, lodging reservations are essential. Don't wait until the last minute!

4. What observing program do you want to conduct? The remaining chapters of this book should give you plenty of suggestions. Plan your program carefully and make a list of the equipment you will need to carry it out. Be frugal; if you try to do everything, you'll likely accomplish very little. You will certainly have an opportunity to witness all of the eclipse phenomena possible from your site, but there is rarely enough time to make systematic and accurate observations and photographs of everything (see **Sperling's Eight–Second Law** in Chapter 9)! Isolate a few projects which have special interest for you, and plan to complete these with the greatest care. If you finish them, you might turn to a few back–up projects. If you are part of a team, you may have to sacrifice a few favorite projects for the good of the group, but a team which shares results and images later is likely to accomplish much more than an individual alone could hope to do.

5. Decide on pre– and post–eclipse sightseeing; you may

Trials Of An Expedition Leader

If you think leading an eclipse expedition is as easy as falling out of bed, think again! A well–rounded leader must be knowledgeable in astronomy, have the psychology of a bartender and the patience of a saint. Over the years I've lead ten solar eclipse expeditions from the Andes mountains to remote taiga of Siberia. One must not only understand the dynamics of the particular eclipse but the country and its customs, geography and weather patterns. Often it's the human factor that proves to be the most challenging. A group of one hundred eclipse chasers will give you as many different personalities who react in as many different ways.

If there ever existed a classic example of the so–called 'love–hate' relationship it must surely be one particular couple who seemed to join every expedition I lead. They delighted in fighting with each other. He loved eclipses—she hated them! No expedition leader wanted any part of this pair. One could sense that beneath that pained shell of a man was a heart that longed for the awesome peace and beauty of a few precious moments of totality.

Nothing seemed to please her; she found fault with everyone and everything. During a luncheon banquet outside Cairo following a successful expedition to Kenya for the 16 February 1980 total solar eclipse she complained that the black pepper on her salad dressing had 'moved.' "This salad is tainted with varmints," she shouted! No one ate their salads. The Egyptian chef was furious but the waiter could have cared less. "I'll sue," she screamed. "Be my guest, Madame," the waiter replied with a smirk.

The best was yet to come. That evening we had planned a gala event to celebrate the expedition's success. At dusk we would head off into the desert to an oasis where we would be treated to a lavish banquet complete with traditional belly dancers! Each eclipse chaser was given the option to ride in a dune buggy or a "romantic" camel ride. My favorite couple began to bicker over the mode of transportation; if he was going to ride a camel then so was she!

I tried to discourage her. "Camels have bad breath. They drool! Please take a dune buggy; it's much safer." She wouldn't listen, her husband was riding a camel and now the crowd was encouraging her. Her camel, named 'Coke,' had been avoided by others because Coke appeared to be moody and unpredictable. Coke knelt down so she could mount. As Coke began to swiftly rise I shouted "hold on tight!" It was too late; Coke already had enough of this woman. As she lunged forward, unable to keep her grasp, she shouted "whiplash . . . I'll sue!"

Everyone roared uncontrollably, especially the camel herder. Someone shouted "Allah Akabah" (God be praised). "See?" I said, the Egyptians are happy." "Egyptians," she retorted angrily, "that was my husband!"

Remember: the human factor is probably the most unpredictable factor in leading an eclipse expedition!

Donald F. Trombino
Deltona, Florida

need advance tickets or reservations for some activities. After all, if you're going there anyway, you may as well play the tourist and see some of the sights!

Chronicle: Open a journal, notebook or computer log to chronicle your eclipse plans from the very beginning. Include a checklist of inquiry letters you write, things you order, and things you need to do; don't assume that just because you've sent out a letter to someone that it will be answered without some follow–up effort on your part (or that you'll remember you wrote it in the first place, if you don't get a reply). Include all pertinent details of your planning, an essential ingredient if you hope to use the journal as a guide for future expeditions. After the eclipse, the journal makes an excellent place to record observations and impressions of the event and other activities you engaged in on the trip.

Localize: You know where you're going; now you need to figure out how you'll get around once you're there. If you don't have a firm observing site lined up already, an open area can usually be secured with the permission of a friendly land owner, but it is vital that the utmost caution be exercised to insure that no private property is damaged, disturbed, or an owner inconvenienced in any way. If your lodging and observing sites are in different places, you'll need transportation and either good maps or a guide. Don't rely on lining these up when you arrive if you can help it. If you haven't seen your site before, you may encounter unpleasant surprises which require you to relocate after arrival. Again, transportation is important. Bad weather might set in. It is unrealistic to expect to get through to a local weather station or to rely on radio or television weather broadcasts on eclipse day. Make arrangements to call a weather service well outside the eclipse path if conditions appear to be deteriorating. Select a location where, if the weather turns bad, there are convenient routes and modes of transportation to get you to other sites along the eclipse path. Will you need—and be able—to rent an automobile or hire a driver for eclipse day? You might even want to consider chartering a private aircraft to get above the weather, which usually requires advance reservations and guaranteed payments.

Instrument: You'll need to instrument your expedition in accordance with your observing program. The further you'll be traveling, the lighter you'll want to travel. Take only the optical and photographic

Photograph 4-1 Smaller instruments not only provide adequate eclipse images but are easier to transport to eclipses. Jackie McDuffie and the 60 mm equatorially–mounted refractor she used to observe the 11 July 1991 total solar eclipse south of Puebla, Mexico. *Photograph taken by Rusty Harvin.*

equipment you'll need, and use a checklist. Consult it every time you pack and unpack. Don't forget accessories, especially filters and photographic or recording supplies. In some locations, popular film, video tape, audio tape and battery brands are unavailable; carry enough for your program needs with you. Take along bad–weather protection for your equipment, especially critical for electronic components such as lap–top computers, tape recorders and telescope drive correctors. Zip lock bags are good for film, eyepieces and other small items; trash bags make good water–tight telescope and camera covers. An important piece of equipment is a spare tape recorder with earphones and pre–recorded eclipse countdown and narration tape. This often overlooked tool can serve as a verbal reminder of what events are upcoming during the eclipse, and what observations or photographs you want to make. Begin the tape and narration ten minutes before totality and continue it

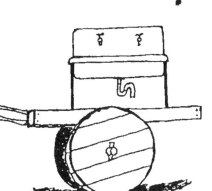

until ten minutes after; this should be ample time to include all major events and your voice cues. Include time advisories on your narration tape; thirty-second intervals should suffice until totality is at hand, and thereafter, ten-second advisories leading into and out of mid-totality are recommended. Record your own observations on a second tape recorder; if you include on this one a comment about when you started your countdown-narration tape, you can coordinate the two later and compare your notes with your expectations. Don't forget short-wave radios capable of receiving broadcast time signals if your program requires them, and remember to list non-astronomical needs, such as artist's brushes, sketchpads, canvas and paints if you have artistic goals.

Practice: This stage should begin as early as practical, but at least several months before the eclipse. The best approach for your first practice is to pack everything you expect to need to fulfill your goals, then carry it out in the field and rehearse most of your program, including photography, using the uneclipsed sun. Trying out your camera and film combinations, exposure times and video taping system can confirm that your equipment is sound and your program feasible for the partial phases at least. There is no better way to discover what you've overlooked or not considered, and you'll have plenty of time to correct any problems. Keep practicing, at least once every couple of weeks, until the date of your departure. Then, when you arrive at your staging area, practice again. Strange things happen to equipment in transit.

Sustain: SCUBA divers have a saying. *"Plan your dive, then dive your plan."* It's good advice for eclipse observers too. You've gone through the trouble of establishing, instrumenting and practicing what you want to do; you have a good plan. Once you're on site, sustain that plan by following through on it. Some flexibility is admirable, but resisting those last-minute doubts and temptations to significantly alter your plans is the best path to success.

Evaluate: The eclipse is over; things may or may not have gone well. Even the best laid plans can often encounter insurmountable obstacles. Most likely, you've enjoyed some successes and experienced some failures. Immediately after the eclipse is the best time to record your reactions and initial evaluations, while they're fresh in your mind. Later, when film is developed and video or audio tape examined, you may have other input. These are the important things that will contribute to your journal being complete and your next eclipse having an even greater likelihood of being trouble-free!

"Remember, exposed film and optics first!"

Solar Eclipse Observing - The Partial Phases

Introduction. The partial phases of a total, annular–total or annular eclipse of the sun, as well as partial solar eclipses themselves, offer a number of interesting projects that can be conducted with modest equipment.

Contact timings. Accurate timings of the moon's contacts with the solar limb may be carried out by pinhole projection, eyepiece projection with binoculars or telescopes, or direct viewing through safe solar filters with naked–eye, binoculars, camera lenses or telescopes.

First contact is the instant when the moon's limb is initially seen encroaching on the solar disk. This is the most difficult contact to time, since you must anticipate where the moon will first appear against the sun's limb. *Fourth contact*, when the moon leaves the sun, is easier to time since the observer can anticipate the event with considerably greater certainty.

The best timing method is to tape record your voice comments over short–wave radio time signals. From North America, the National Institute of Standards and Technology broadcasts time signals over radio station WWV from Fort Collins, Colorado (WWVH from Hawaii) on frequencies of 2.5, 5, 10, 15, and 20 MHz. Similar signals may be received on frequencies of 3.330, 7.335, and 14.670 MHz over radio station CHU from Ottawa, Ontario, Canada. An increasingly popular alternative method is to video tape the sun's actual or projected limb while the camcorder records the radio time signals as audio. Subsequent playbacks, with the aid of a stopwatch, can effectively eliminate timing errors brought on by failure to visually spot the encroaching moon at the earliest possible moment.

If you are unable to receive radio time signals in the field, a wristwatch, checked for accuracy shortly before and as soon after the eclipse as possible, will suffice. Be certain to record, along with your timings, an estimate of their accuracy (i.e., "to the nearest 0.1 minute") for later reference.

In addition to noticing differences in the precision with which you time first contact and fourth contact, you should also notice a significant difference in timings done by observers at the same site using different types of equipment. An interesting study is to compile and graph data on various observers' timings of each contact against their methods and/or the diameters of the instruments they used. Such studies must be conducted with some care, however, to prevent observers located too closely together from influencing one another's judgments as to when contacts are actually seen.

Sunspot contact timings. Timings of contacts of the moon's limb with individual sunspots or sunspot groups are especially interesting. Observers with access to short–wave radios can make fairly accurate timings of such contacts. Others, having stopwatches or wristwatches, can accurately time how long it takes for a sunspot or group to be covered.

The angular velocity of the moon in its orbit about the earth averages 33 minutes of arc (written 33') per hour, or 0.55' per minute of time. Since the sun's angular diameter averages 32' the moon should appear to sweep across the sun's visible disk in 58 minutes. Since the sun's diameter is 1,391,000 km (864,400 mi), in order to cover that diameter in 58 minutes, the moon must sweep across some 400 km (257 mi) of the sun's face each second. Of course, this blatantly neglects the substantial effect of the sun's limb curvature. Still, for sunspots close to the center of the solar disk and for a lunar passage that is nearly central, it is possible to make reasonable estimates of the diameters of sunspots and sunspot groups from the number of seconds it takes the moon's limb to obscure them.

Limb irregularities. Some observers have recorded seeing

Photograph 5-1 An annular eclipse of the sun is in many respects a partial eclipse. Annular eclipse of 24 December 1973 from Cruz Verde, Colombia, photographed with clouds as the filter. *Photograph taken by Richard Sweetsir.*

irregularities or deformities on the lunar limb as it moves across the sun. Whether such irregularities represent actual mountains and valleys on the lunar surface, illusions caused by the contrast between the moon and the sun, or simply atmospheric disturbances is debatable. However, they are worthy of note, especially if their appearance times and their actual compass position angles or their locations with respect to nearby sunspots are recorded for later comparison with observations by others.

Photographic projects. Partial eclipses lend themselves to several interesting photographic effects. There are, of course, the obvious targets already described. Frequent and accurately timed exposures around first and fourth contacts make impressive sequences allowing contact timings to be extrapolated. High-quality camcorder video, with radio time-signals on the audio track, can be reduced later to yield precise contact timings. Photographing the stages of lunar encounters with sunspots and attempting to image any limb irregularities should also be tried. But there are other worthwhile photographic projects that have no observational counterparts.

A camera capable of multiple exposures on a single frame may be centered on the sun's expected position at mid-eclipse (keep in mind that the sun moves about 15 degrees each hour). If multiple exposures are taken at intervals of five to ten minutes, the resulting photograph will show the progression of the partial eclipse stages across the frame. Practicing this technique a few weeks before the eclipse will insure the best combinations of exposure interval, shutter speed, f-stop and lens for the film you've selected.

A similar effect can be obtained with video camcorders capable of shooting time-lapse sequences; if exposures are frequent enough, the sun will appear to sweep across the sky while the eclipsed portion first grows then shrinks. Mounting the camcorder on a clock-driven telescope mount will freeze the sun's motion across the sky while animating the eclipse stages, allowing use of telephoto lenses.

Still cameras may also be used to take time exposures of the sun as it crosses the sky. The technique is similar to that used to produce star trails, but requires denser filtering than normal exposures would. The resultant streak of sunlight should decrease in width as mid-eclipse approaches, then increase again as the moon moves off the sun's disk. For annular or total eclipses, you might want to interrupt the exposure around mid-eclipse, and snap a normal exposure of the eclipsed sun in the center of the streak path (remembering to appropriately reduce the filtering for an annular or remove it altogether for a total eclipse). Again, practice in advance of the eclipse!

Photograph 5-2 Binoculars are excellent for eyepiece projection as demonstrated by Jeremy Reynolds at the 10 May 1994 annular eclipse. Note the white cardboard at the objectives producing a shadow at the projected image.

Finally, don't neglect still and video photography of yourself, other observers around you, your site and the various equipment set-ups. Eyepiece projection of the partial phases or of annularity is as effective on someone's T-shirt as it is on a projection screen; be creative! Such images add much to the overall eclipse experience.

Specific exposure recommendations for photographing partial solar eclipses may be found in Chapter 17 and video imaging in Chapter 18.

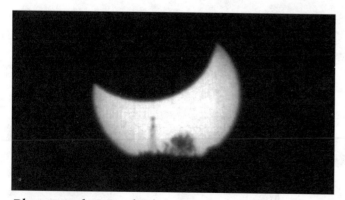
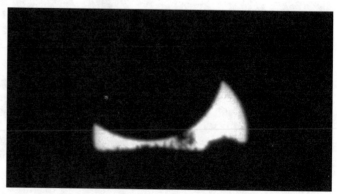

Photograph 5-3a, b The partial eclipse at sunset following the total solar eclipse of 12 October 1977 as seen from Colombia. Note the transmission tower (left photograph) and the tree. *Photographs taken by Carter Roberts.*

Solar Eclipse Observing - Environmental Studies

Introduction. In the fifteen-or-so minutes on either side of totality or annularity, subtle changes take place in the environment and in the behaviors of wildlife. Family members without extensive astronomical equipment or projects to conduct can make valuable contributions as observers of their physical and biological surroundings.

Pinhole camera images. As the sun dwindles to a thin crescent and the ground lighting dims, hundreds of tiny images of the crescent sun appear projected among the shadows cast by leaves beneath their trees. Nature produces this fascinating phenomenon in much the same fashion as observers who make pinhole camera viewing devices; in this case, the pinholes are the tiny spaces between the densely packed leaves.

A worthwhile experiment for even young children would be to try out a variety of manufactured and home-made devices to determine what apertures produce the clearest and sharpest pinhole images. Some recommended household items are colanders, sifters, strainers, squares of pegboard and perforated convection-oven trays. Even the fingers of their own hands, spread slightly apart, will project these crescent shapes. They might also be alert to the types of trees and leaves, or other foliage, which produce the best natural images.

Diminishing light. As the sun dwindles to a smaller and smaller crescent, the surrounding landscape very gradually takes on a late-afternoon appearance. In the final minutes before totality, however, the changes in lighting become quite striking, and the skylight seems to diminish in stepped increments rather than with a smooth, gradual transition toward darkness.

The changing colors of the surrounding landscape and distant clouds can be quite striking as well, with yellows and oranges most often described. Nor do light intensity and color changes necessarily mirror themselves coming out of totality. Many describe the sun's return as abrupt and the colors less pronounced or of different hues following totality, while others believe these perceptions to be more psychological in origin; certainly, further investigation is warranted.

Artists and poets have a distinct advantage over photographers and technocrats in capturing diminishing-light phenomena and the feelings that accompany them.

Wildlife behaviors. Animals are very perceptive to changes in their environment. Their responses to the unexpected early nightfall and even more unexpected daybreak after such a short night, can make a fascinating study.

Livestock, wild birds, squirrels, insects and even domesticated pets will behave in unusual and interesting ways because of the environmental changes that accompany a total or dark annular solar eclipse. Some animals, accustomed to feeding schedules dictated by dawn and dusk, exhibit changes in eating habits on eclipse day and, according to some reports, for several days after an eclipse. Others, roosters, for example, crow on cue as twilight comes and goes, and appear to experience no disorientation at all.

Mosquitoes have been known to go on feeding frenzies, and even aquatic life is not spared; fish are reportedly more willing to bite around totality (if one can imagine taking time out from such a celestial spectacle to go fishing), and their neighboring amphibians have been heard to carry on quite loudly.

Studies of solar eclipse effects on individual forms of wildlife, except for occasional anecdotal accounts of diminished milk production by cows and egg production by chickens, are few at best. People with interests in these areas are encouraged to pursue them.

Meteorological observations. Most eclipse-chasers have at least a basic familiarity with meteorology and weather forecasting. Many amateur astronomers carry along meteorological instruments to measure temperature, wind, air pressure and humidity changes throughout an eclipse. For others, a primary reason for observing the eclipse may be meteorology.

There are significant changes in the weather during eclipses. As the amount of sunlight is reduced, the temperature of the surrounding air begins to fall. This results in corresponding changes in the barometric pressure, wind speed and direction, dew point and humidity. These changes can sometimes result in unexpected fog or dew formation that can hinder eclipse observations and photography; they should be anticipated and equipment should especially be monitored for dew formation if the relative humidity is high.

A simple grade-school experiment can quickly determine the dew point at an observing site prior to totality. Slowly add a few shards of ice, a little at a time, to a glass or metal container of water. The moment a thin mist of dew forms on the outside of the container, insert a thermometer into the water. The water temperature at that moment is the dew point of the surrounding air, or the temperature at which dew and fog will form. If the air temperature is sufficiently higher, you should have no problem with dew; if it is within five degrees or so, it would be wise to monitor your equipment carefully.

Recording readings from meteorological instruments are good projects for younger family members or those without astronomical observations to conduct, and are excellent additions to your eclipse journal. Instruments to consider taking are thermometers, barometers, hygrometers, anemometers (simple hand–held wind–speed devices are available from most science supply houses), and compasses for determining wind direction.

If your site is near an airport or television station, consider making prior arrangements with the resident meteorologists to allow you to stop by after the eclipse and photograph their chart recorders or copy their data for the various weather elements. Some will also have instruments which record sunlight intensity throughout eclipse day. An interested researcher or astronomy group might wish to query weather stations all along the eclipse path to obtain comparative data.

Weather satellite imagery. Many schools and amateur meteorologists have access to inexpensive groundstation equipment capable of receiving real–time U.S. NOAA and Russian Meteor satellite images on their personal computers. Local HAM radio clubs or weather stations may be able to put you in contact with them.

If a satellite pass coincides with the time of totality, the moon's shadow on the earth's surface (or on cloud tops) should be visible in the image received. Stored in their computers as TIFF (Tag Image File Format) files, they may be easily copied onto 3.5–inch high–density microdisks and transferred to your own hard drives for viewing later. Service bureaus, desktop publishing companies and photo processors in major metropolitan areas can provide photographic prints from these computer files.

The original NOAA images, incidentally, contain both visible and infrared frames side–by–side; the latter may be processed, through the system's software, to provide remote temperature readings within the moon's shadow.

Following the eclipse, weather satellite imagery of the moon's shadow might also be found on computer bulletin board systems (BBS's) as GIF (Graphics Interchange Format, pronounced JIF) files that may be downloaded by modem to your own hard drive or high–density microdisk and viewed with the appropriate software.

Most computer desktop publishing and painting software supports TIFF files (identified by the suffix .TIF). GIF files (.GIF) save and retrieve slower than TIFF files, but use an efficient compression algorithm which keeps file size small and saves disk space, making them a favorite for BBS's and for more rapid uploading and downloading.

Photographic projects. A light–colored surface, such as the side of a house, a street, sandy ground, bed sheet, or piece of white cardboard, will make the leaf–formed pinhole camera images of crescent suns stand out for photography. Still or video available–light photographic

Photograph 6-1 GOES weather satellite image of the 11 July 1991 total solar eclipse. Note the moon's shadow west of Baja California, Mexico.

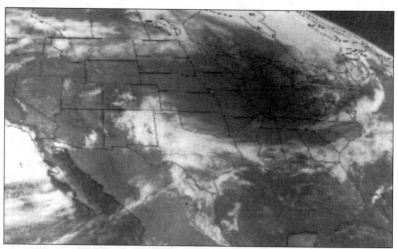

Photograph 6-2 GOES weather satellite image of the 10 May 1994 annular eclipse. Note the moon's shadow over the midwestern states. Also note the lack of sharpness when compared to the moon's shadow in **Photograph 6-1**.

sequences, at fixed aperture and shutter speed, would be useful in determining when these images are first and last visible on each side of totality or annularity.

Photography of diminishing light effects may be attempted, especially with low–light video cameras. Also, don't neglect photographic studies of wildlife behaviors and photo documentation of chart recordings from weather instruments.

Chapter 17 offers exposure recommendations for photography and Chapter 18 gives advice for video imaging.

Solar Eclipse Observing - Shadow Bands

Introduction. For a few minutes on either side of totality, when the sun is a thin crescent and the sky takes on the characteristic yellowish tint of eclipse–induced twilight, alert observers may notice eerie bands of undulating shadow racing across the ground, along the sides of buildings or across other light–colored surfaces. This atmospheric phenomenon, first described by H. Goldschmidt in 1820 but undoubtedly observed since antiquity, is believed caused by an irregular bending or refraction of the crescent sunlight.

Explanations. Attempts to explain the *shadow band* phenomenon have focused upon the atmospheric rather than the astronomical sciences, for the bands do not reliably appear or exhibit similar behaviors at each eclipse, even when eclipses have similar geometric circumstances.

In 1900, H. C. Wilson proposed a diffraction–ring hypothesis for their origin. Writing in the now–defunct magazine **Popular Astronomy**, Wilson maintained that a pattern of concentric rings surrounded the moon's shadow, and the shadow band movements were caused by this pattern moving across the earth.

In the November, 1925, issue of that same magazine, U.S. Weather Bureau meteorologist W. J. Humphreys reported that the bands paralleled the solar crescent and moved normally to their length during an eclipse earlier that year. He concluded that the bands were *pseudo–total reflections, or mirage effects produced by transition shells between warmer and cooler adjacent masses of air in a state of thermal convection,* and that there was no relation between their direction of travel and that of the winds aloft at any level.

In the June, 1963, issue of **Sky & Telescope** magazine, Edgar Paulton of the Amateur Astronomers Association of New York City suggested their movements in such widely divergent directions, when viewed from different locations, were an illusion brought on by their rapid passage.

Many observers hold strongly to the belief that changes in shadow band direction are caused by variations in atmospheric temperature, humidity, density and pressure, contending, therefore, that it would be difficult at best to predict their appearance and motions for any specific eclipse in advance.

To obtain the fullest possible picture of these enigmatic bands, observations of their presence (or absence) and behaviors, along with relevant meteorological data, are desired for all total and near–total annular eclipses. Observations are also desired from all possible sites along the length and width of each eclipse path, in order to obtain comparative data for various geometries and weather conditions. Only then will a definitive explanation for shadow bands and a reliable means of predicting their likelihood of occurrence for any given eclipse be possible.

Appearance. Shadow bands vary considerably in both width and separation, but range most frequently between 2 and 5 cm (0.75 and 2 in) in width and are separated from one another by 5 to 25 cm (2 to 10 in).

Their direction of motion across the ground seems to depend upon where an observer is located along the eclipse path and whether the bands are observed before or after totality. Their velocities vary most often between 1.5 and 3 m (5 and 10 ft) per second.

Equipment. Casual observers of shadow bands can detect them against any light–colored surface. For serious studies of the bands, however, a standardized observing procedure was developed by Paulton in 1959 to simplify comparisons of observations made from widely separated sites.

Paulton's approach calls for setting up projection screens perpendicular to the axis of the shadow cone. Each observer is then examining a tiny segment of a plane upon which the moon's shadow appears circular.

First, a portable wooden frame about 1.5 m (5 ft) square is constructed and supported by diagonal struts joined at the center. The finished frame and mount should resemble an oversized artist's easel.

Next, the screen surface is selected. This may be a white

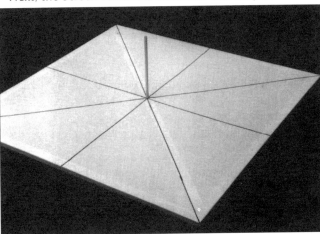

Photograph 7-1 A simple shadow band screen.

cloth, sheet or canvas which completely covers the frame (eclipse-chaser Norm Sperling recommends sandpaper for greater contrast). A large azimuth circle, marked off and labeled in five-degree increments, is then drawn or painted onto the screen's face. This is best done by projecting a slide or overhead transparency of an azimuth circle onto the screen and tracing the image.

Finally, a rod is mounted in the center of the screen, projecting about 30 cm (12 in) out of the middle of the azimuth circle.

If made of light-weight materials and designed for easy assembly and breakdown, this piece of home-made equipment can be transported and set up with a minimum of effort.

Photograph 7-2 Bill Riebsame, left, inspects a shadow band screen prior to the 7 March 1970 total solar eclipse while Karl Simmons and Murray Daw, right, look on. *Photograph taken by Karl Simmons.*

Procedures. Paulton's ideal shadow band observing team requires three people. A few minutes before totality, the screen is oriented so that the central rod casts no shadow.

When the first bands appear, the first observer stretches a length of string across the screen parallel to the length of the bands. The string may be attached to the screen with push-pins, thumb-tacks, or tape. Some have designed a more elaborate frame with hooks attached to the screen at each five-degree mark on the azimuth circle; the string is then wound around opposing hooks.

If the bands do not change direction, no further action is necessary. If they do, every 30 seconds another string is attached. The strings should be prepared in advance, labeled in numerical order with masking tape flags and laid out in the order in which they are to be used.

The first observer must also notice the direction of motion of the bands, which may or may not parallel their length. A twelve-hour clock face system is recommended to avoid confusion with the orientation angles on the screen. The observer might, for instance, observe the orientation of the bands as stretching between 225 degrees and 45 degrees, while their direction of motion is from 11 o'clock to 5 o'clock.

The second observer, armed with a stopwatch, alerts the third observer with a pre-arranged countdown and starts the stopwatch the moment one of the bands begins its journey across the screen. At that same moment, the third observer begins counting the number of bands which cross some predetermined position on the screen, usually the edge from which the bands are leaving.

Meanwhile, the second observer has been following that one shadow band across the screen. The moment the selected band leaves the screen, the second observer simultaneously stops the stopwatch and alerts observer three with another verbal signal. Observer three, upon hearing observer two's outcry, stops counting the departing shadow bands and records the number counted.

When finished, the first observer will have recorded valuable data on the orientation and direction of travel of the bands, the second observer the time of travel across the screen, and the third observer the number of bands present in the circle's diameter at that time.

Paulton has recommended that measurements be repeated at 30-second intervals for as long as shadow bands are visible, and that the entire experiment be repeated following totality. If frequent observations are planned, tape-recording the various measurements, directions and counts, preferably against a background of WWV or CHU broadcast time signals, can minimize confusion later.

Paulton's experiment is relatively simple and well suited to clubs or school classes which could be formed into several teams and located at different sites. Considerable practice by prospective team members is required to guarantee success, however.

Observations from totality's edge. Observers located near the edges of the path of totality will experience shadow bands for longer periods of time. However, since there is no guarantee that a particular eclipse will produce the bands, and since the duration of totality diminishes rapidly as the northern or southern limit of an eclipse path is approached, observers will sacrifice valuable minutes from the most spectacular part of a total eclipse. Large groups planning to send volunteer teams to these hinterlands should make certain that the participants fully understand what they'll be giving up in totality time to obtain more information on this interesting phenomenon.

Photography and Video Imaging. Dim lighting, poor contrast and rapid movement make still photography of shadow bands difficult at best. Low-light camcorders offer a significant advantage, and can be centered on the viewing screen to capture the motions of the bands for later playback and data reduction. Specific recommendations may be found in Chapters 17 and 18.

References

Paulton, Edgar M. *Observing and Reporting Shadow Bands.* **Sky & Telescope** (September, 1959). Sky Publishing Corporation, 1959.
Paulton, Edgar. *Eclipse Shadow Band Motions—An Illusion?* **Sky & Telescope** (June, 1963). Sky Publishing Corporation, 1963.

Solar Eclipse Observing - The Lunar Shadow and Sky Darkness

Introduction. As the moment of totality approaches, the shadow of the moon sweeps out of the west to immerse observers along its path. This spectacular sight, often compared with that of an approaching thunderstorm, is even visible—indeed, sometimes enhanced—when cloudy weather threatens to obscure totality itself. With the arrival of totality or a particularly deep annularity, the sky darkens appreciably and the brighter stars and planets become visible.

Appearance. Noted portrait painter Howard Russell Butler described the shadow's appearance through thin cloud cover over the Elkhorn Range and intervening valley at Baker, Oregon for the eclipse of 8 June 1918:

...a greenish pallor overspread the landscape,—but it was not very dark. To the northwest, however, the sky was growing dark. The last half minute seemed long. My eyes were fixed on the sky line. Suddenly the entire range fell to a deep low-valued blue, and simultaneously the lower part of the sky above the range turned to a rich yellow, inclining to orange, streaked with two horizontal blue-gray clouds. Above me the sky darkened rapidly. For an instant the valley retained its light green color and then the shadow seemed to rush toward us and all was engulfed...

Overcast skies disappointed unfortunate observers viewing the 7 March 1970 total eclipse from Florida's panhandle. As consolation, however, the heavy cloud cover provided an impressive canvas for an especially striking apparition of the rolling, wave-like shadow as it swept out of the Gulf of Mexico and across the Southeastern United States.

The appearance of the shadow varies considerably depending upon the circumstances of the eclipse, atmospheric conditions and the location of the observer with respect to the center line. For instance, a sunrise eclipse will cast a shadow resembling a truncated cone with its narrow end pointing towards the horizon, while observers near the northern or southern limits of the path find themselves surrounded by asymmetrically illuminated horizons.

Even observers several hundred miles outside of the path of totality have reported sightings of the shadow sweeping along the horizon. Certainly, observers who, for one reason or another, find themselves close to the path of totality but unable to travel the remaining distance, would want to attempt shadow sightings along with their observations of the partial eclipse. Brilliant red and orange hues (the sunrise–sunset effect) appear as horizon colors prior to, during and just after totality.

Once the shadow's leading edge has swept over your site, the sky darkens significantly. Observers who have adapted their vision with dark goggles prior to the onset of totality have reported bright stars and planets becoming visible more than a minute before the thin solar crescent is obscured.

Similar sightings of celestial objects have been reported in the much brighter skies of annular eclipses since antiquity; observers are encouraged to include such attempted sightings in their observing programs for annular and very deep partial eclipses as well as for total eclipses of the sun.

Procedures. In the 1970's William H. Glenn, York College of the City University of New York at Jamaica, conducted and compiled extensive observations on the appearance of the shadow. He developed a questionnaire which still provides one of the best sets of guidelines for lunar shadow and sky darkness observations. Among the major shadow and sky aspects to note, Glenn includes:

1. How many seconds before totality was the shadow seen?
2. In what direction did the shadow become visible?
3. How did its darkness, appearance and color compare with the surrounding unshadowed sky?
4. How would you describe its very rapid motion and changing appearance, color and shape?
5. In what directions did you have time to look during as well as immediately before and after totality?
6. How would you describe the appearance and color of the sky in each of these directions as totality progressed?
7. When and how did the shadow disappear?
8. Describe how the light changed as the eclipse progressed (include any times as accurately as possible).
9. What stars and planets were you able to identify?
10. How soon before totality did stars and planets become visible?
11. How long after the end of totality did stars and planets disappear?
12. What kind of weather conditions existed during the eclipse (include the fraction of sky obscured by clouds and the direction of clouds, types of clouds, cloud coloration, sky coloration, haze, etc.)?
13. Did you detect any color on the disk of the moon during totality?

14. How did the darkness and color of the moon's disk compare to the sky a few degrees away from the eclipsed sun?

Observers should also estimate the degree of darkness using such criteria as the readability of newspaper headlines, ordinary newsprint, the hour or second hand of an ordinary wrist watch, a standard eye chart or some similar method. Measurements made with photoelectric cells or photographic light meters would be of even greater value. Use devices capable of measuring the incident light on a scale that can be readily converted to lumens per square meter (or foot–candles), and make frequent carefully–timed readings of the intensity of illumination at the zenith and at the four compass points along the horizon.

Locating even the brightest stars and planets prior to totality can be difficult unless you know exactly where to look and have taken some steps to dark–adapt your vision in advance. This task has been greatly simplified by widely published charts of the sky around the eclipsed sun. Such charts, which appear in most periodicals and astronomical handbooks for total and annular eclipses and a number of excellent computer programs that produce such charts (see Chapter 2), can be very useful in estimating the faintest stellar magnitude visible. This, in turn, provides yet another strong indicator of sky darkness.

Selecting a star or planet you have identified during totality, and attempting to keep it in sight in the rapidly brightening sky following totality, is an easier chore.

Photography. While capturing the essence of the lunar shadow is a task especially suited to artists and poets, photographers having cameras equipped with fish–eye lenses (a 180–degree field is ideal) have enjoyed considerable success. Mount your camera on a level tripod, point it at the zenith and orient it with respect to the compass points (corrected for any difference between the earth's magnetic and geographic poles at the observing site), and each frame should capture the entire sky from horizon to horizon, showing the sunrise–sunset effect and darkness as the shadow advances.

Glenn achieved good results by leaving the lens wide open (about f/5.6) and making one–second exposures every 15 seconds from a minute before until a minute after totality on a moderately fast film (around ISO 200). This will yield overexposed images at first, but as the shadow sweeps over, exposures become ideal and produce a sequence of images that is both pleasing and useful. Frames taken during totality frequently show brighter stars and planets as well.

With the compass points penned onto each photograph, it is a simple matter to determine the azimuth of the approaching and receding shadow, and to relate any sunrise–sunset effect noticed to an accurate position on the horizon. The exact time of each exposure can be determined later if a tape recorder is used to simultaneously record the triggering of the camera shutter and WWV or CHU radio time signals (see Chapter 5).

Video cameras with wide–field lenses or attachments should produce impressive results as well, but should be adjusted for full manual operation to prevent them from automatically correcting for the diminishing light by adjusting exposure times or f/–stops or both throughout the eclipse.

Chapters 17 and 18 contains further information on exposures for shadow photography and video imaging.

Photograph 8-1 A 165° fish–eye lens was used to capture the lunar shadow, easily seen in this 1/8 second exposure on Ektachrome 100 of the 26 February 1979 total solar eclipse north of Winnipeg, Manitoba. *Photograph taken by Mike Reynolds.*

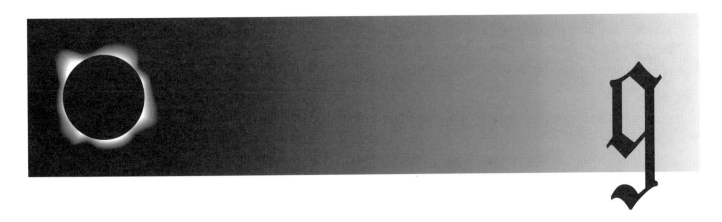

Solar Eclipse Observing - The Diamond Ring and Baily's Beads

Introduction. Two spectacular events signal the boundaries of totality for seasoned eclipse observers, the appearances of the *diamond ring* effect and *Baily's beads*. The beads are also an annular phenomenon.

Unlike Baily's beads, the diamond ring, for all its spectacle, is not a true phenomenon of totality but a product of the final moments of the pre–totality partial phases and their post–totality resurgence.

Sir Edmund Halley is credited with making the first observations of Baily's beads during the eclipse of 22 April 1715. They were also seen by Maclaurin from Edinburgh during the annular eclipse of 1 March 1737 and by Williams from Revolutionary War America on 27 October 1780 (see Chapter 1) from just outside the path of totality. But it was Francis Baily's widely–disseminated description of the phenomenon during the annular eclipse of 15 May 1836 that led to their bearing his name thereafter.

Explanations. Shortly before second contact of a total eclipse, the opposing horns of the slender crescent sun begin to converge on one another. At the same time, the tenuous solar atmosphere becomes visible against the darkening sky, shining out around the edge of the moon where the sun has already been covered. The combination of this "ring" of light and the single brilliant "diamond" of sunlight where the horns are converging creates a most striking appearance, the diamond ring.

Photograph 9-1 Francis Baily. *Portrait photograph © The Royal Astronomical Society.*

The effect lasts for a very short time. Soon, the horns of the solar crescent close completely, and the diamond begins to break up, to be replaced by an array of brilliant beads of sunlight caused by the sun shining through valleys and depressions on the moon's leading limb.

Baily's beads also quickly succumb to the encroaching moon, winking out one or two at a time until totality is fulfilled; the disappearance of the last bead marks the

Photograph 9-2 A diamond ring of the 11 July 1991 total solar eclipse from Cabo San Lucas, Mexico. This exposure was taken through an 80 mm f/8 refractor at prime focus with a Nikon F3 camera body on Ektar 125 film. *Photograph taken by Doug Berger.*

moment of second contact and the beginning of totality. Totality ends with third contact when the beads reappear at the opposite (trailing) limb. Shortly thereafter, another diamond ring appears and quickly absorbs the beads as the horns of the crescent sun diverge once more.

In an annular eclipse, Baily's beads first appear along the sun's trailing, rather than leading limb, pinpointing the moon's deepest limb depressions first and marking second contact. The beads increase in number as the limb moves further onto the sun's disk until no breaks in the sun's rim are visible. When the moon's leading limb reaches the opposite rim of the sun, the beads return, then diminish in number as

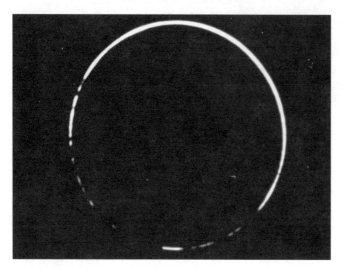

Photograph 9-3 A photograph of a projected image, 30 May 1984 total–annular eclipse, Atlanta, Georgia, showing Baily's beads. *Photograph taken by Richard Sweetsir.*

they merge together until the last one vanishes when the moon's limb appears unbroken, marking third contact.

Procedures. Diamond ring and Baily's beads observations are limited primarily to measurements of their position angles, which are measured from north through east around the moon's limb. This may be accomplished either by direct observation with appropriate filters, by photography, or by eyepiece projection onto a screen.

In both cases, care must be taken to insure proper orientation of the sun's image with respect to the compass points. The best approach here is to use either an illuminated eyepiece reticle marked off in degrees or a similar scale drawn on the projection screen. Proper orientation of the reticle or screen may be accomplished by careful monitoring of the westward solar drift during the partial phases.

The last, faint beads visible around second contact, and the first ones visible around third contact of a total eclipse, or first and last beads, respectively, of an annular eclipse, have the greatest scientific value and should be carefully timed, measured and photographed.

It is possible to prolong the diamond ring and Baily's beads phenomena by setting up an observing station near the edge of the path of totality or annularity. Observers at such locations sacrifice totality or annularity time, which is significantly reduced near the northern and southern path limits, but carefully timed measurements of the beads made during such "grazing" eclipses have greater scientific value and can yield fairly precise measurements of the shapes and relative positions of the sun and moon. A leading proponent of this work has been Dr. David W. Dunham and the International Occultation Timing Association. Dunham has spearheaded related research through observations of grazing occultations of stars by the moon.

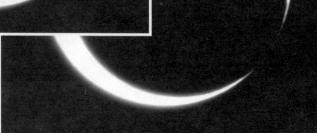

Photograph 9-4a–c

A series of photographs taken at the northern limit of the 5 January 1992 annular eclipse, Truk Lagoon, showing the development of Baily's beads. The first photo was taken at 21:22:00 UT, middle photo at 21:23:14 UT and the right photo at 21:23:17 UT. These 1/125 exposures were taken through a 1000 mm f/11 lens on Kodachrome 64. *Photographs taken by Derald D. Nye.*

Sperling's Eight–Second Law

Everyone who sees a total solar eclipse remembers it forever. It overwhelms the senses, and the soul as well. Then it hits you: "It was supposed to last a few minutes—but that couldn't have been true. It only seemed to last eight seconds!"

This effect frustrated my first four eclipses. Yet tape recordings, movies, and the whole edifice of celestial mechanics all claim that it did last the full, advertised two to seven minutes.

The culprit is attention span. If you stare transfixed, your mind, knowing the scene isn't changing, says "I already know that," and neglects to store away the same image yet again. So the solution is not to stare.

Pre–record a tape cassette, timed to start at the first diamond ring. On it, tell yourself what to notice during different parts of your precious few minutes in the moon's shadow. Notice how the umbra envelops you, enjoy the diamond ring, then examine the prominences (they're bright so you don't have to be fully dark–adapted). Next, survey the corona—its general shape, and outstanding features.

Switch away for a few seconds, to check the colors all 360 degrees around the horizon. Since totality is just starting, it will be darkest in the west, lighter in the east. Now, back to the sun. Your eyes, now partly dark–adapted, are ready for the corona. Which is the longest streamer, and how far out can you trace it? Where is the innermost dark wedge? Pick out an interesting pattern of filaments and make a mental engraving of it.

OK, back to the horizon. Sweep around again, and notice how much difference a minute or two makes. The west is lightening, foretelling totality's end, and the east is dark, where folks down–path are just now getting theirs.

Finally, back to the sun. Review the most noteworthy coronal details. Look again at prominences, since there's a whole new crop of them on the 3rd–contact side. Watch for the pink fringe of chromosphere that anticipates—yes, here it comes—the second diamond ring.

How quickly the corona fades—and now even the last of it is going—and it's incredible how bright even that tiny wedge of the sun's surface can be!

And now this eclipse, too, is over—but this time you have won. From each separate span of attention during totality you can savor your eight seconds of mental replay. If you moved your attention enough times, you'll recall many times that eight–second limit. Yes, Sperling's Eight–Second Law can be beaten!

Norman Sperling
Oakland, California

Photography. The diamond ring and Baily's beads should be photographed without the dense solar filters used for visual observing, yet they are still partial eclipse phenomena and should be considered hazardous to one's eyesight. The safest approach is to center the camera on the sun during the last seconds of the partial phases while the solar filter is still attached.

When the horns of the crescent converge to form the diamond, remove the solar filter and begin exposures without looking through the viewfinder again. Specific exposure recommendations may be found in Chapter 17, but you'll want to bracket them somewhat in order to capture the full range of ring and bead appearance. But keep in mind that if you're shooting around f/11 at 1/250 second (or its equivalent) with ISO 200 film to capture the ring and beads, you will not want to photograph totality at these settings! Remember to adjust your exposure settings accordingly once the beads are gone and totality begins (and again at the end of totality when you are shooting the beads and ring again).

Video photography may be attempted with camcorders in the automatic exposure mode and the solar filter removed for satisfactory results, but most photographers will prefer to set exposure parameters manually here too. Vary the f/–stop through a pre–determined range to accent beads of varying intensities. Unlike still and motion picture cameras, most video camcorders have viewfinders which are miniature television monitors and present no danger to the eyesight since you are not looking at the sun itself. Still, safe habits are good to develop, especially if you are using an assortment of visual and photographic equipment, and exposure adjustments are easy to make without looking through the viewfinder. If you have one of the earliest models of a home video camera, consult your manual carefully for warnings about "burn–out" before pointing it directly into the sun!

It may be interesting to note that Baily's beads were first successfully photographed at the eclipse of 7 August 1869 by C. F. Hines and members of the Philadelphia Photographic Corps from a site in Ottumwa, Iowa.

Viewing dangers. These are still considered partial eclipse phenomena and great care must be taken when observing the diamond ring or Baily's beads. Direct viewing, even through a camera viewfinder, requires use of adequate solar filters safe for visual use (see Chapter 3).

Solar Eclipse Observing
- Second and Third Contacts,
- Flash Spectrum,
- Chromosphere, and
- Prominences

Introduction. Totality or annularity arrives suddenly, and for totality, with striking spectacle. *Second contact* marks the beginning of totality or annularity. For annularity, this occurs when the moon's trailing limb reaches the sun's disk. For totality, it is when the moon's leading edge completely covers the sun's disk. Totality is further identified as the moment when Baily's last bead winks out, or when the brightness of the sun's *photosphere* fades to the level of the adjacent *chromosphere* and inner *corona*.

The *flash spectrum* is seen at second contact, when the photosphere's dark absorption lines are suddenly replaced by the chromosphere's bright emission lines. The chromosphere itself, a beautiful pink/red to scarlet color, is visible near second and third contacts.

During totality, jets of glowing gas, ranging from red to scarlet, may be seen arching more than 30,000 km (19,000 mi) from the sun's disk. These are the *prominences*, most commonly found in the higher solar latitudes and in greatest number a few years after a sunspot minimum.

Finally, *third contact* and the reappearance of Baily's beads ends totality or annularity. For annularity, this occurs when the moon's leading limb leaves the sun's disk. For totality, it is when the moon's trailing limb exposes the sun's photosphere once more.

Contact timings. Timings of the second and third contacts are considered important and should be carried out for all central eclipses; different procedures are recommended depending upon which contact is being timed and whether the eclipse is total or annular.

Eyepiece projection onto a white screen offers the best method of timing second contact for total eclipses, since the photosphere projects prominently while the chromosphere does not. Visual observations by unaided eye or binoculars, using safe visual filters, yield adequate results for general purposes. However, the brightening of the chromosphere and corona as totality draws near, and the lack of a clear-cut separation between the photosphere and the chromosphere, make visual measurements somewhat less reliable than those made by eyepiece projection.

For annular eclipses both methods seem to work equally well, although the higher resolution (i.e., the ability to discern fine detail) afforded by binoculars or telescopes makes it easier to determine the instant when the last tiny segment of the moon's dark limb moves onto the sun's face and the annulus becomes complete.

During annular eclipses a few observers have reported a "black drop" irradiation effect (where the planet's limb seems to delay its separation from or hasten its reunion with the sun's limb) similar to those observed at solar transits of the planets Mercury and Venus, where that last lunar limb segment appears to linger on the edge of the sun, elongate, then separate suddenly. Watch for it, as it can complicate contact timings.

Third contact is best timed by the reappearance of Baily's beads for a total eclipse. The first bead's appearance stands out in stark contrast to the dimmer phenomena of totality, and presents little difficulty for eyepiece projection or direct viewing methods.

For an annular eclipse, an optical system providing the best available resolution is preferred. Observers should watch for any "black drop" effect at third contact as well.

Photograph 10-1 Second contact of the 10 May 1994 annular eclipse from El Paso, Texas. SVHS image taken through a 68 mm lens with a 5X telextender. *Video taken by John Westfall.*

Flash spectrum. The flash spectrum was anticipated,

first observed, and named such by Princeton professor C. A. Young at the total eclipse of 22 December 1870, who described it in this way:

...the moment the sun is hidden, through the whole length of the spectrum, in the red, the green, the violet, the bright lines flash out by hundreds and thousands, almost startlingly; as suddenly as stars from a bursting rockethead, and as evanescent, for the whole thing is over in two or three seconds.

Pogson was first to observe the flash spectrum during an annular eclipse on 6 June 1872.

Inexpensive diffraction gratings and spectroscopes are widely available and amateurs with an interest in more technical observations are encouraged to include the flash spectrum in their programs.

Chromosphere. This innermost region of the sun's atmosphere is primarily of historical significance, since it was in the spectrum of this light during an eclipse in 1868 that the element helium was first identified more than 25 years before it was finally recognized on earth. The chromosphere exhibits the same red to scarlet color as the prominences, and should be identified and noted as a part of any total eclipse observing program, but it is essentially featureless and warrants only passing attention.

Prominences. What the chromosphere lacks in structure, the prominences make up for. Stannyan first described prominences in a letter to Flamsteed following the eclipse of 1706, but the first detailed descriptions of them were by the Swedish astronomer Vassinius at the eclipse of 1733 (although he incorrectly believed them to be lunar in origin). Spanish admiral Ulloa, observing the eclipse of 24 June 1778 from sea, suggested they were caused by sunlight shining through breaks in the moon's limb, but they were not widely accepted as a solar phenomenon until the eclipse of 1842.

Prominences may be relatively quiescent, persisting for many weeks without significant change, or they may be violently active, erupting outward as far as three million kilometers (two million miles) from the sun's surface. The more active prominences will exhibit changes from minute to minute; noting any movement or change is a worthy undertaking.

Observers viewing with telescopes equipped with eyepiece reticles (system of lines or dots in the focus of an eyepiece) might wish to measure the position angles of the prominences arrayed about the sun's limb. Techniques for doing so may be found in Chapter 9.

The extent of particularly broad prominences, especially those which loop about and return arch-like to the solar surface, may be estimated in a similar manner. Some reticles have a vertical extension scale as well as an azimuth circle; these may be used to measure the relative heights of prominences above the sun's limb.

Photography. Still and video sequences around the time of second and third contacts can pinpoint the times of these events when taken in conjunction with tape–recorded time signals; video cameras are especially advantageous here.

Flash spectrum spectroscopy utilizing diffraction gratings and simple spectroscopes can be especially rewarding. Try the wide selection of fast color films and low–light video cameras on the market and practice by attempting to capture spectra of the sun and the full moon to determine the best system for you. Howerver note that flash spectrum photography is very challenging!

Specific exposure recommendations for the chromosphere and prominences may be found in Chapter 17, but it is wise to bracket exposures somewhat to capture the varied structure exhibited by the prominences.

Viewing dangers. Great care should be taken when timing second and third contacts to prevent eye damage. Safe visual filters are required whenever the sun's photosphere is visible. Flash spectrum observers using diffraction gratings should be careful to look only at an angle through the grating and not directly at the sun. The chromosphere and prominences do not present any danger from infrared radiation, and may be viewed without filters.

Photograph 10-2 Prominences visible at the 11 July 1991 total solar eclipse from Cabo San Lucas, Mexico. Note the distinctive 'seahorse' prominence at the bottom (south). This exposure was taken through an 80 mm f/8 refractor at prime focus with a Nikon F3 camera body on Ektar 125 film. *Photograph taken by Doug Berger.*

Solar Eclipse Observing - The Corona

Introduction. As totality or very deep annularity arrives, the pearly-white solar atmosphere flashes into view surrounding the new moon's darkened disk. Known by such colorful names as "halo" and "glory" from earliest times, this is the corona, a tenuous region of free electrons and interplanetary dust extending millions of kilometers into space and shining in the eclipse sky with the intensity of the full moon.

Mention of the corona appears in literature dating back to a description of the 20 March 71 solar eclipse in Plutarch's dialogue **On the Face in the Orb of the Moon**. It was not until the eclipses of 8 July 1842 and 28 July 1851 that the corona was positively associated with the sun instead of the moon.

The inner or K-corona, composed largely of electrons, displays a continuous spectrum and provides a backdrop for viewing prominences. The outer dust or F-corona has an absorption spectrum, extends for several solar diameters, varies in shape from eclipse to eclipse but is generally irregular, and features spectacular radial streamers.

Procedures for observing the corona should focus on its shape and extent, intensity variations and streamers.

Shape and extent. The sun's corona varies from a generally uniform and circular appearance around times of sunspot maximum to a highly irregular one near sunspot minimum.

Measuring the shape and extent of the sun's corona is a project well-suited for the observer without special optical equipment. This may be done descriptively with the aid of a tape recorder (and hand-held reticle for estimating position angle), or artistically using rough sketches or "mapping" techniques.

Many science, engineering and surveying supply houses sell inexpensive glass reticles etched with bull's-eyes, linear scales and azimuth circles. An excellent home-made combination bull's-eye or linear scale and azimuth circle may be drawn, then photographed or photocopied onto transparency film. When mounted in a cardboard viewing tube, it is a simple matter to sight on the eclipsed sun and orient the azimuth circle's "zero degree" position with the sun's north pole. The corona's extent may then be called out and tape recorded in terms of solar radii at chosen position angle intervals.

Experienced sketch artists may accomplish the same thing by pre-sketching a landscape feature close to the eclipsed sun, then using the feature for scale and orientation when roughing in the corona's shape and extent during totality.

Figure 11-1 Totality presents a wonderful opportunity for artists to represent the event, often in more detail than film or video. *Artwork by Jeanne Pusateri.*

Those having less artistic skill may wish to attempt a rough diagramming technique known as mapping, where a circle to represent the eclipsed sun is drawn in the center of a sheet of graph paper, and the surrounding squares are shaded in during totality to match the appearance of the corona.

Intensity variations. The corona does not exhibit the same brightness intensity throughout. These subtle differences, which occur both radially and circumferentially with respect to the sun, are caused by variations in the density of the coronal gases and tend to "wash out" in photographs. However, using techniques similar to those employed for shape and extent observations, the uneven texture of the corona's brightness may be readily described or sketched.

Descriptions may be recorded in terms of position angle and extent using a five-step linear scale, where 0 is the darkest (the appearance of the moon's center at mid-eclipse, or of the surrounding sky) and 5 is the most brilliant part of the corona.

Sketch artists will want to employ a heavier stroke to emphasize brighter regions, while mappers might use the same 0–5 intensity scale to outline changes within the squares on their graph paper in a "paint by numbers" approach.

Streamers. These long finger–like strands near equatorial regions and short fan–shaped appendages near the poles of the sun align themselves with the sun's magnetic field. They also vary significantly in number, structure and distribution throughout a sunspot cycle.

Estimates of their number, along with descriptions of the shapes, position angles and extents of the more prominent streamers, might be undertaken.

Sketch artists, especially those equipped with binoculars or low–power telescopes, might want to make detailed renderings of the more spectacular streamers.

Photography. The corona—in fact, totality itself—was photographed successfully for the first time by Berkowski at the eclipse of 28 July 1851; using the 15.9 cm (6.25 in) Königsberg heliometer, he required an exposure of 24 seconds. Fortunately, modern photographic emulsions offer much greater versatility.

With any diffuse subject like the corona, its appearance on film is a function of the exposure time and f/–stop used. There is wide latitude and ample time to shoot a sequence at different settings, from the shortest exposure likely to produce an image to time exposures as long as 15 seconds for moderately fast films (around ISO 200).

Shorter exposures will capture greater structural detail in the inner corona, enhancing the streamers, while failing to capture the impressive extent of the outer corona; longer exposures do justice to the outer corona, while washing out the other details. Skilled darkroom or computer technicians can combine individual exposures to create composite images which capture the best of both extremes.

Video photography should be attempted with automatic focus and exposure features disengaged, since they prove highly unreliable under low lighting conditions. Experiment in advance with the full moon, and settle on a range of f/–stops and focal lengths which show detail on lunar surface features. This should provide an adequate mid–range for coronal photography with most camcorders.

See Chapter 17 for specific information on photographing the corona and Chapter 18 for video recommendations.

Viewing dangers. During totality there is no danger to the eyesight from solar radiation. It is safe to view the corona and coronal streamers without special filters, even through optical devices. However, it is crucial that care be taken to discontinue any direct visual viewing before third contact, when the sun's disk is again exposed to view. A countdown tape with built–in several–second margin of error and attention–getting signal, or a timekeeper serving the same purpose, will provide a reliable warning system.

Photograph 11-2 The inner corona of the 26 February 1979 total solar eclipse. *Photograph taken by Jay Anderson.*

Photograph 11-3a, b These photographs of the 11 July 1991 corona were taken normally (left) and with a vignetting disk (right). Note the greater detail in the inner corona visible with the vignetter, while the extent of the streamers is almost the same in both images. The defocusing of the vignetted image was anticipated for this experimental photograph but is not the fault of the vignetting disk; proper focus is achieved by simply focusing with the glass vignetter–support in the optical system. *Photographs © by Stephen J. Edberg.*

Solar Eclipse Observing - The Sky at Totality

Introduction. Shortly before second contact, the sky darkens appreciably and the brighter stars and planets become visible to the unaided eye. With the onset of totality, stars fainter than the second magnitude may be seen.

Searches for faint comets, asteroids or infra–Mercurian planets have been carried out during total eclipses for more than a century, but usually such searches, to be fruitful, require sophisticated cameras which produce images for later analyses.

Total eclipses even provided the first experimental support for Einstein's General Theory of Relativity. The theory predicted that light from a distant star would be bent by the gravitational field of the sun (due to the curvature of space itself) and the star would appear deflected slightly from its actual position in space. Just such a deflection was detected in a star close to the sun's limb during the total eclipse of 29 May 1919, only four years after Einstein presented his theory.

Observers who are able to tear themselves away from the beauty of the eclipse itself, may find some interesting and worthwhile targets among the celestial objects which

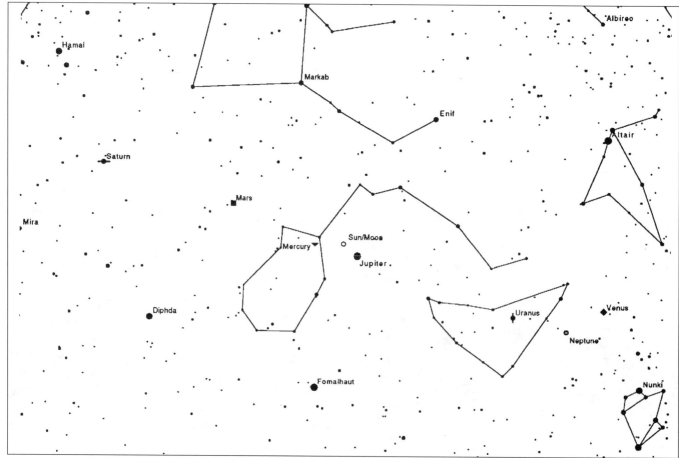

Figure 12-1 The position of stars and planets relative to the totally–eclipsed sun of 26 February 1998. Popular astronomical magazines publish such diagrams in advance of an eclipse. In addition many astronomy computer programs create such illustrations. *Illustration courtesy of Voyager II for Macintosh.*

would otherwise be hidden from view by the sun for another month or two.

Popular astronomical magazines, handbooks, almanacs and computer planetarium programs usually provide diagrams of the sky during totality, including the brighter objects likely to be visible.

Some observers want to dark–adapt their eyes well in advance of totality, and do so by wearing dark glasses or a single eye patch during the partial phases.

Planets. The inner planets, Mercury and Venus, are rarely far from the sun, and are usually observed close to the horizon through turbulent layers of atmosphere and in twilight. During totality, however, both planets may be viewed high in the sky. Planetary enthusiasts might take a few minutes from a particularly long eclipse to observe and sketch telescopically the dusky markings which Mercury and Venus exhibit.

Outer planets Mars, Jupiter and Saturn, when near solar conjunction, may offer out–of–season targets for enterprising planetary astronomers wishing to see what features they are exhibiting in advance of their return to morning skies from the sun's glare.

Variable stars. Variable star enthusiasts may wish to check on the brighter irregular or eruptive stars on their programs. Certainly, any novae discovered shortly before becoming lost in the sun's glare would warrant examination during totality.

Comets. Observations of bright comets in close proximity to the sun might be conducted. Others may wish to search for previously undiscovered comets which would otherwise remain lost in the sun's glare for weeks more. Dividing the sky around the sun into sections assigned to many observers would be more likely to achieve success than would one observer attempting to sweep a large area of sky during the brief period of totality. Use wide–field telescopes, and avoid areas known to have a lot of bright nebulae or galaxies.

Meteors and fireballs. An alert observer who spots a bright meteor or fireball might make an effort to record its time, magnitude, duration and direction of flight. For very bright objects, plotting the path on a star chart or "Sky at Totality" illustration would be useful. Note any bursts (and associated sounds, if any) and any lasting trains left behind by its passage.

Artificial satellites. The brighter satellites should be readily visible during totality. They are inarguably of less relevance than planets, variable stars or comets, which may be obscured by the sun's glare for many weeks around conjunction. Still, the widely available computer software and modem–accessible elements warrants running predictions for the eclipse site and an effort to observe them. This task is ideal for people with few other projects to conduct.

Faintest star. One project which relates to the sky's overall darkness during totality is to determine what the faintest star visible to the unaided eye is. This may also be attempted using binoculars and telescopes of various apertures. The best approach is to select a star which you know will be bright enough to locate easily and which also is surrounded by fainter stars offering a broad range of magnitudes. Different observers might focus on several different star fields at increasing distances from the eclipsed sun to determine how sky darkness varies as the shadow's edge is approached. Observers who have not dark–adapted at least one eye in advance should not expect their faintest star observations to have any value.

Zodiacal light. The zodiacal light is a faint glow that travels along the ecliptic preceding and following the sun by up to several hours, and appears to be caused by sunlight reflecting off interplanetary dust particles in the inner solar system. In the evening and morning hours, it may appear as bright as the Milky Way at times when it stands nearly vertically off the horizons. For the western evening sky, this occurs in February and March for mid–northern latitude3, while the eastern morning sky is favored in September and October, providing there is no moonlight to interfere.

The sun's brightness makes it difficult to study the size and density of these dust particles closer than about 20 degrees (80 solar radii) from the sun. However, some studies of the zodiacal light have been conducted closer to the sun during eclipses using cameras carried aloft by aircraft and balloons. University of Minnesota scientist Dr. Edward P. Ney led several expeditions to areas where the eclipsed sun would be ten degrees below the horizon, thereby enhancing the zodiacal light.

The zodiacal light is probably too faint to be observed visually during totality, except from very high elevations. However, eclipse observers with a very low sun angle, or who find themselves just outside of totality for an eclipse occuring before sunrise or after sunset at their locations, might wish to attempt observations of this nature.

Gegenschein. This "counterglow" phenomenon is related to the zodiacal light but appears on the ecliptic directly opposite the sun. It appears as a very faint elliptical patch of light. Observers who find themselves directly opposite some point along the path of totality might monitor the gegenschein for several nights on either side of the eclipse date and several times centered on the time of totality in their opposite hemisphere for any changes in its appearance.

Photography. Still, video and CCD imagery of the brighter planets should present no obstacles. Short exposures on moderately fast films through wide–field lenses should capture the appearance of the sky at totality with the eclipsed sun as the centerpiece. Bright satellites and meteors should also be relatively easy targets, provided exposures are kept short enough to prevent sky fogging; practice on full–moon nights or at twilight with exposures ranging from a few seconds to a minute. The obstacles presented by most variable stars, novae, comets and the zodiacal light, especially under the illumination and during the brief time afforded by totality, are probably too formidable for those without highly sophisticated imaging equipment.

Reference

Reynolds, Mike. *Some Solar Eclipse Highlights*. **Sky & Telescope** (Sept. 1973). Sky Publishing Corporation, 1973.

13

Lunar Eclipse Observing – The Penumbral Eclipse

Introduction. When a full moon occurs close enough to the ascending or descending node of the moon's orbit, but not sufficiently close to make contact with the earth's dark umbral shadow, a penumbral eclipse takes place.

Unless the moon's limb passes within about 1120 km (700 mi) of the umbra, corresponding to a penumbral magnitude of about 0.500, the unaided eye is unlikely to detect any light reduction. However, there are a number of interesting observations which may be undertaken by observers of deep penumbral eclipses and for the penumbral phases of partial and total lunar eclipses.

Detection timings. Timings of first and last penumbral shadow detection are impossible, as there is no clear–cut easily discernable shadow edge to be seen. Instead, the penumbra appears as a somewhat dusky darkening on the limb, which advances steadily across the surface.

Determining when you believe you see this darkening is highly subjective and can vary greatly depending upon whether observations are made by telescope, binoculars or unaided eye. Don't expect to detect the penumbra visually until at least 30 minutes (and possibly as long as 45 minutes) into the penumbral phase of a partial or total eclipse. For shallow penumbral eclipses, the shadow may go undetected throughout the eclipse. In making penumbral timings observers are encouraged to estimate their probable accuracy as well.

Some observers tape record a continuous narrative of the moon's appearance, including times when contacts or any dimmings are first suspected and finally certain. Others have reported some advantage by racking the moon's image slightly out of focus to blur the lunar surface features, which vary significantly in albedo (reflectivity), making the moon's brightness more uniform.

Shadow position. Contact timings should also include an indication of the position along the lunar limb where the shadow is first detected. This may be done by position angle, using either an eyepiece reticle or lunar map, or by reference to a known feature near the lunar limb.

Umbral separation. Penumbral observations during eclipses which will be partial or total should include estimates of the greatest distance the penumbra may be detected from the umbra. Estimate the separation for both the leading and trailing penumbral shadow. This is easier to determine than penumbral detection timings, and can provide a reasonable check on detection timing reliability. Use a linear or bull's–eye eyepiece reticle, or note the positions of the umbra and penumbra with respect to lunar features, then measure their angular or actual separation from a lunar map.

The use of neutral density or paired polarizing filters to significantly dim the moon may be helpful in detecting the penumbra; color filters may be used to enhance the shadow as well.

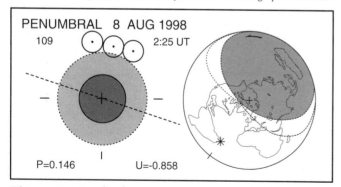
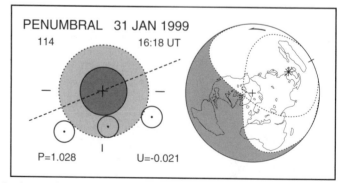

Figure 13-1a, b These figures represent the moon's path through the earth's penumbral shadow (the outer, lighter–shaded circle) as well as visibility of the eclipse on the earth (no eclipse visible in shaded area). The 8 August 1998 penumbral eclipse is very shallow; little if anything will be visible. The 31 January 1999 penumbral eclipse is quite deep; that is the moon passes very deeply within the earth's penumbral shadow and barely misses being a partial lunar eclipse. Such penumbral eclipses have a better chance of being observed. *Illustrations by Fred Espenak, NASA Goddard Space Flight Center.*

Shading variations. Penumbral shading is not uniform; it becomes darker the closer it gets to the umbra. Care should be taken not to confuse variations in the penumbral shadow with differences in the albedos of lunar surface features. Photocopies of simple lunar maps may be used to shade in variations.

Color variations. While not as spectacular as umbral colors, the deeper penumbra can present a variety of dusky brown or yellowish–brown colors. Note variations on photocopies of lunar maps; numbers might be assigned to different colors and written inside sketched boundary lines on maps. Use of a variety of color filters at the eyepiece is recommended.

Photoelectric photometry and intensity variations. Penumbral theory is not well established. There is a distortion in the solar disk, provided by the intrusion of the earth's atmosphere. This results in wavelength–dependent differences in the illumination of the penumbra from what would be expected from an earth having no atmosphere. These differences are not completely understood, making photometry of features within the penumbra a worthwhile undertaking (see Chapters 14 and 15 also).

The brightness of different parts of the penumbra can be estimated in terms of a graduated scale based upon the brightness of different lunar features, for those lacking photoelectric equipment.

Other observations. The penumbral eclipse or penumbral stages of a partial or total eclipse, when the full moon's brightness is diminished, are good times to look for lunar transient phenomena (LTP); lunar surface materials may exhibit fluorescence, caused by solar wind interactions. Chapter 14 discusses LTP observing in greater detail and recommends specific lunar features for special attention.

Timing when you can last and first see shadows of lunar features on the limb opposite the penumbral shadow's center is another interesting project to conduct.

Photography. Frequent exposures offer a possible alternative to visual estimates for contact timings. Using slow films (around ISO 32) and deliberately underexposing them, the full moon can be dimmed sufficiently on initial exposures to show a detectable change some 30 minutes after the penumbra begins its advance (note, however, that this method will not give pleasing images of the penumbral eclipse itself). Shoot at six–second intervals, beginning around 30 minutes after the predicted time of penumbral contact, using an exposure previously determined to barely show the full moon's disk with your system. Be certain to finish and change to a faster film and exposure setting well before first umbral contact!

A similar approach might be employed with video cameras equipped with time–lapse features which can produce a speeded–up eclipse sequence to accent the penumbral advance.

Guidelines for photographing the penumbral eclipse are provided in Chapter 17 and video in Chapter 18.

Photograph 13-2 Penumbral eclipses can often be recorded only through the use of a photometer. This is a simple photocell photometer for penumbra brightness measurements. *Photograph taken by Francis Graham.*

Reference

Graham, Francis G. and Westfall, John E. **Lunar Eclipse Handbook**. East Pittsburgh: Lunar Press, 1990.

Lunar Eclipse Observing - The Partial Phases

Introduction. Partial lunar eclipses occur when the moon enters the earth's umbral shadow, but not centrally enough to become completely immersed. This chapter suggests projects which are suitable for partial eclipses and the partial phases of total lunar eclipses.

Program management. The partial phases are busier than totality for lunar eclipses; as busy as the time around totality for solar eclipses. If your observing program is particularly packed, you may want to consider pre–recording a carefully–timed countdown tape which directs your attention to each of the things you want to do during the eclipse. Produce the tape from a script with a timeline having a built–in 15– to 30–second safety margin. These have proven very helpful at solar eclipses (see Chapter 4: Instrument), but it's a good idea to make a backup copy or carry the original script along if you're going to rely completely on this (or any) device.

Tape–recording observations and timings has become a popular alternative to frequent note–taking in the field, but it is important to remember that tapes can run out or jam, and recorder batteries can fail; check your equipment frequently to avoid losing valuable data!

Contact timings. First contact, when the earth's umbra first touches the moon's celestial eastern (lunar western) or leading limb, and fourth contact, when it last departs the moon's celestial western (lunar eastern) or trailing limb, should be timed to an accuracy of 0.1 minute (6 seconds). Neutral–density filters are recommended to enhance the umbra's appearance if glare is a problem.

Instruments having apertures between 10 and 40 cm (4 and 16 in) and very low magnifications (between 40 and 100X) are preferred for contact timings, although smaller telescopes, binoculars and even the naked–eye may be used. The important consideration is that most or all of the moon fit into your field of view at one time. The darker penumbral shadow is often dark enough to merge with the umbra when large–aperture instruments are used at high power, complicating accurate timings; beginners should

Photograph 14-1 A partial eclipse of the moon is a spectacular sight even if a total eclipse will not be visible. *Photograph taken by Mike Reynolds.*

especially be alerted to this possibility.

Recording your verbal descriptions against WWV or CHU radio time signals is the most reliable method of obtaining good timings. Begin your narrative several minutes prior to the predicted times of the two contacts.

If you are without a short–wave radio, you might be able

Plate 1 left: Partial phase of the 10 May 1994 annular eclipse from Las Cruces, New Mexico. *Photograph taken by Don Trombino.*

Plate 2 below: Multiple exposure of the 10 May 1994 annular solar eclipse from El Paso, Texas. These 1/125 second exposures were taken over a 30 minute period using a 200 mm, f/11 telephoto on Ektar 125 film. *Photograph taken by Ewell Schirmer and James Engelbrecht.*

Plate 3a, b above and below: The 10 May 1994 annular eclipse photographed through a Thousand Oaks visual glass filter and an 80 mm refractor on Ektachrome 200. *Photographs taken by Mike Reynolds.*

Plate 4 right: The 4 January 1992 sunset annular eclipse from San Diego, California taken through a 400 mm telephoto and a 2x telextender on Fuji Velvia ISO 50 at 1/30 second exposure. *Photograph taken by Alan Gorski.*

Plate 5 right: The flash spectrum at third contact of the 11 July 1991 total solar eclipse. The zero order direct image is to the left, first order spectrum is centered, and the more

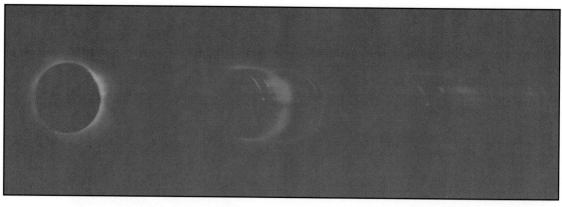

dispersed, fainter second order spectrum is on the right. In the first order spectrum the sharp arcs with knots at their bottoms are due to Hydrogen α in red, an unresolved mix of Sodium D_1, D_2, and Helium D_3 in orange, the Magnesium β line group faintly in green, Hydrogen β in light blue, and a very weak Hydrogen γ in blue. Between the Hβ and Hγ the red Hα image of a prominence on the eastern limb is visible. The fuzzy red and green circles are the coronal Red line due to Iron X and the coronal Green line due to Iron XIV, Iron atoms have lost 9 and 13 electrons (of 26), respectively. *Photograph © by Stephen J. Edberg.*

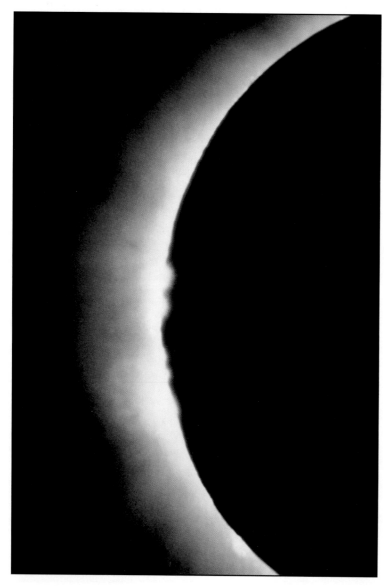

Plate 6 above: The horizon and sunrise–sunset effect of the 1988 total solar eclipse from Bangka Island, Indonesia, taken on Kodachrome 200 through a 50 mm lens. *Photograph taken by Carter Roberts.*

Plate 7 left: The 16 February 1980 total solar eclipse from Kenya. A 1/500 second exposure on Ektachrome 200 through a 200 mm Schmidt–Cassegrain. Note Baily's beads, prominences and the corona. *Photograph taken by Alan Gorski.*

Plate 8 above: The 11 July 1991 total solar eclipse from Cabo San Lucas, Mexico. An 80 mm f/8 fluorite refractor at prime focus with a Nikon F3 body and Ektar ISO 125 film was used to capture the coronal details. *Photograph taken by Doug Berger.*

Plate 9 right: The 3 November 1994 total solar eclipse from Iguaçu Falls, Brazil. A 70–210 mm f/3.5–4.0 zoom telephoto at 210 mm f/4 was used to take this 1/60 second exposure on Fuji RHP 400. *Photograph taken by Volkmar Schorcht.*

Plate 10 below: The 11 July 1991 total solar eclipse from Baja California, Mexico. An exposure was taken every six minutes on specially–cut Fuji Velvia through a Kodak Recomar and a 135 mm lens. *Photograph taken by Carter Roberts.*

Plate 11 right: The 30 December 1982 total lunar eclipse and background stars. Two minute exposure through a 1000 mm f/10 lens on Fuji 400. *Photograph taken by John Westfall.*

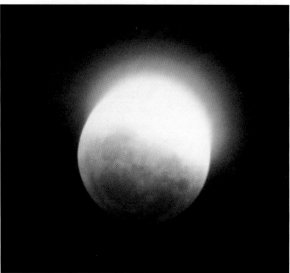

Plate 12 above: The total lunar eclipse of 6 September 1979 through an 800 mm f/12.6 lens. This ten minute exposure was taken on Ektachrome 400. *Photograph taken by Conrad Jung.*

Plate 13 right: On orbit! The Moon rising in eclipse, 17 October 1986 at 1800 UT, as observed by the GOES satellite. *Imaging by Jay Anderson.*

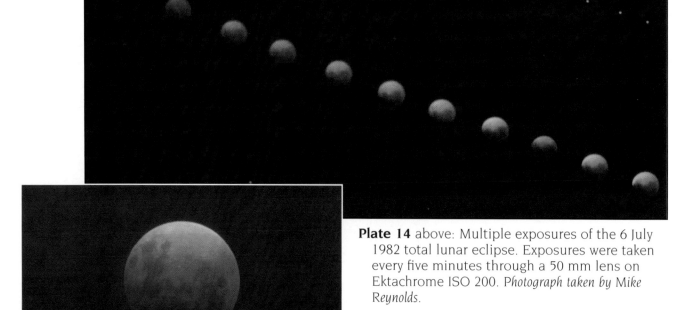

Plate 14 above: Multiple exposures of the 6 July 1982 total lunar eclipse. Exposures were taken every five minutes through a 50 mm lens on Ektachrome ISO 200. *Photograph taken by Mike Reynolds.*

Plate 15 left: The 24 April 1986 total lunar eclipse. A 64 second exposure was taken through a 1000 mm f/11 lens on Ektachrome ISO 200. *Photograph taken by John Westfall.*

to calibrate your timepiece to another's radio before and after the eclipse, and apply an average correction to your timings if necessary. WWV also offers a long distance telephone time service; the number is (303) 499–7111, but as this signal is being transmitted through telephone lines, only assume accuracy to a second or so.

If none of these timing methods are available to you, time the interval between first and fourth contact as accurately as possible with a sweep second–hand, digital wristwatch or a stopwatch.

Photograph 14-2 The twenty craters recommended for lunar eclipse contact timings by the Association of Lunar and Planetary Observers. See **Table 14-1**, below, for identification of craters 1 through 20. *Illustration by John Westfall, A.L.P.O. Lunar Section. Lick Observatory Photograph.*

Table 14-1 Recommended Craters for Lunar Eclipse Contact Timings									
1.	Grimaldi	6.	Timocharis	11.	Manilius	16.	Gassendi	21.	Billy
2.	Aristarchus	7.	Tycho	12.	Menelaus	17.	Birt	22.	Campanus
3.	Kepler	8.	Plato	13.	Plinius	18.	Abulfeda E	23.	Dionysius
4.	Copernicus	9.	Aristoteles	14.	Taruntius	19.	Nicolai A	24.	Goclenius
5.	Pytheas	10.	Eudoxus	15.	Proclus	20.	Stevinus A	25.	Langrenus

You should not anticipate or allow your observations to be unduly influenced by the predicted contact times. The predictions are determined by the diameter of the earth's umbra. However, due to the earth's thick atmosphere, the umbral diameter is routinely larger by approximately two percent than the simple geometry would expect. Since the umbra's diameter, and even its shape, varies from one eclipse to another, careful contact timings can be instrumental in determining both its diameter and degree of ellipticity.

Crater contact timings. As the umbra sweeps across the full moon's face, its leading edge encounters many surface features. These encounters are easier to time than first and fourth contacts, and are equally capable of yielding information on the size and shape of the umbra.

Timings are best accomplished using small–aperture instruments; larger telescopes should be "stopped down" by covering the front of their tubes with cardboard masks in which off–center circular openings of 8 to 10 cm (3 to 4 in) in diameter have been cut. This will reduce the brightness of the lunar surface sufficiently for reliable crater contact timings. Neutral–density filters may also be employed to further darken the lunar surface and to enhance the umbra. As with contact timings, use voice over radio time signals and aim for an accuracy of 0.1 min (6 sec).

The Association of Lunar and Planetary Observers (A.L.P.O.) recommends 20 craters for special attention during lunar eclipses; the Royal Astronomical Society of Canada (R.A.S.C.) adds five more to that list, although any well–defined craters are fair game. Table 14–1 on the previous page compiles the recommendations of both organizations (craters numbered 21–25 are the R.A.S.C. additions).

Small craters are fairly easy to time, for the umbral shadow engulfs them relatively quickly. For larger craters (i.e., Copernicus and Tycho), time the umbra's arrival at first one rim and then at the opposite rim of the crater; later, average the two timings to determine a mean crater contact time.

Observers who do not observe the moon regularly are encouraged to familiarize themselves with the appearances and locations of these recommended craters on several full–moon nights prior to the eclipse; a good lunar map suitable for field use is an indispensible tool. Consider laminating the map to protect it from dew and to allow for the use of grease pencils or transparency–marking pens.

Umbral characteristics. The general appearance of the umbra varies from one eclipse to the next, and may vary during the course of a single eclipse. Observers should note the sharpness (or diffuseness) of the umbra's leading and trailing edges frequently during the partial phases. Note the time when specific variations are detected, and pinpoint their proximity to lunar features on your map.

Be also attentive to the brightness and color of the umbra's interior, and whether any variations are noted as the eclipse progresses. Standard color charts might be employed for the latter, while the former lends itself well to comparisons with the brightness of lunar mare, rilles or rays on the part of the surface still outside the umbra. Failing in that, an arbitrary numerical scale may be developed at the telescope and calibrated for several familiar features already inside the umbra (0=darkest to 5=brightest works well). The brightness of other features within the shadow can be quickly compared to the calibration features, using intermediate decimal values if significant variations are detected.

Apparent magnitude estimates. While generally reserved for observations of totality, estimates can be made of the moon's apparent magnitude throughout the partial phases by comparing its brightness to the magnitudes of bright surrounding stars and planets.

One approach recommended by the A.L.P.O. is to reduce the apparent size and brightness of the moon with a highly–polished convex reflector. A hemispherical hub–cap from an automobile (old Volkswagen "Beetle" ones are ideal), an silver Christmas–tree ball ornament, or a convex bicycle mirror works well.

View the moon in the reflective surface while viewing several stars of known magnitudes with the unaided eye. Adjust your eye's distance from the reflector (the further the distance the greater the dimming) until the brightness of the reduced lunar image matches that of one of the stars, then measure that distance. The following equation yields the moon's apparent dimming or change in magnitude, ΔM (pronounced "delta–M"), from that of the normal full moon:

$$\Delta M = K - 5 \log R,$$

where R is the distance from the eye to the reflector's surface, and K is a constant which needs to be determined just before or immediately following the eclipse. When ΔM is subtracted from the full moon's normal -12.7 magnitude, the result is the magnitude of the eclipse–dimmed moon. To find K, use the same observing method on the uneclipsed full moon (magnitude -12.7) and an equivalent comparison star, set ΔM equal to their magnitude differ-

---

Example: Shortly before an eclipse begins, the planet Jupiter (magnitude -2.1) appears to equal the brightness of the full moon's image in a bicycle mirror 254 cm (100 in) from your eye (the full moon's magnitude is -12.7). Setting ΔM equal to the magnitude difference between the two objects, we find

$$K = \Delta M + 5 \log 254$$

$$K = -10.6 + 5 (2.40) = 1.4.$$

At maximum umbral phase, the moon's reflected image appears to equal the brightness of a star of magnitude 0.4 when you are 35 cm (14 in) from the mirror. Applying the constant K already determined, we find that

$$\Delta M = 1.4 - 5 \log 35 = -6.3,$$

and the moon's eclipsed magnitude is $-12.7 - (-6.3)$, or -6.4.

ence, and solve the equation for **K**.

A.L.P.O. Eclipse Recorder Francis G. Graham suggests using a pivoted convex mirror and a pivoted plane mirror side–by–side. His "geoumbrascope" device allows the mirrors to be adjusted so that the moon and comparison object are seen together; the math is identical.

Another A.L.P.O. method reduces the moon's size and brightness to simplify comparisons with stars. You view the moon through the objective end of a pair of reversed binoculars with one eye while the other eye, without optical aid, seeks out a suitable comparison star of equal apparent brightness.

The magnification of the binocular is used to determine the amount of dimming in stellar magnitudes. Table 14–2 provides the dimming factors for the most common binocular magnifications.

Table 14-3 Atmospheric Extinction Magnitude Corrections

		\multicolumn{8}{c}{Elevation of Lower Object}

		70°	60°	50°	40°	30°	25°	20°	15°
Elevation of Higher Object	70°	0.0	0.0	0.0	0.1	0.2	0.3	0.4	0.6
	60°	—	0.0	0.0	0.1	0.2	0.3	0.4	0.6
	50°	—	—	0.0	0.1	0.2	0.3	0.4	0.6
	40°	—	—	—	0.0	0.1	0.2	0.3	0.5
	30°	—	—	—	—	0.0	0.1	0.2	0.4
	25°	—	—	—	—	—	0.0	0.1	0.3
	20°	—	—	—	—	—	—	0.0	0.2
	15°	—	—	—	—	—	—	—	0.0

Table 14-2 Reversed–Binocular Magnitude Dimming Factors (F)

6x	=	4.20 mag.	7x	=	4.54 mag.
8x	=	4.83 mag	9x	=	5.08 mag.
10x	=	5.31 mag.	11x	=	5.52 mag.
12x	=	5.71 mag.	16x	=	6.33 mag.

Factors for magnifications not given in Table 14–2 may be calculated from the expression

$$F = 5 \log P + 0.31,$$

where **P** is the "power" (magnification) of the reversed binocular (the constant 0.31 assumes some absorption in the optics).

The magnitude **M** of the moon is readily found from the naked–eye comparison star's magnitude **m** and this factor **F**, by

$$M = m - F.$$

With this method no full–moon calibration is required.

> **Example:** A partially eclipsed moon viewed through reversed 10x binoculars (dimming the moon by 5.3 magnitudes) resembles a star of magnitude 1.3; its magnitude **M** would be $1.3 - 5.3 = -4.0$.

A third A.L.P.O. approach employs two small identical telescopes placed close together. Inexpensive imports from the same manufacturer are ideal. One is focused on the moon while the other is trained on a comparison star and racked out of focus to make the star's image size equal the moon's. When the brightness of a star's unfocused image through one instrument equals that of the moon's focused image in the other, the moon's magnitude equals the star's (one telescope will suffice if you calibrate your setup; look at the moon naked–eye and at the star with the out–of–focus telescope). To be effective for the partial phases, however, the moon must be at least as dim as the brightest comparison planet or star available (the brightest planet, Venus, ranges from –3.9 to –4.7; the brightest star, Sirius, is –1.4). Consult monthly magazine or annual handbook sky calendars for planet magnitudes around the eclipse date; for star magnitudes, annual astronomical handbooks and observing guidebooks are good sources.

If you are nearsighted, you have still another approach available. Remove your glasses and, naked–eye, compare the moon with a bright star!

Finally, unless both the moon and your chosen comparison star are at the same elevation, the magnitudes derived by each of these methods must be corrected for differences in atmospheric extinctions for the two objects. Table 14–3 provides rough atmospheric extinction correction values for the moon's previously–estimated magnitude. Add the correction value if the moon is the higher object in the sky, and subtract it if the moon is the lower object.

Lunar transient phenomena (LTP). During an eclipse, the moon undergoes a significant reduction in the amount of sunlight falling on its surface.

During the penumbral phases, shorter wavelengths of solar radiation bathe the moon's surface and may cause some lunar rocks to fluoresce. During the umbra's passage, the surface is subjected to very sudden changes in temperatures which might cause stresses, strains and surface cracking. Such structural deformations could trigger a mild form of lunar vulcanism, releasing trapped subsurface gases and stirring up lunar dust possibly visible from the earth.

Observers should monitor areas before, during and after umbral passage which have historically been suspected of exhibiting *lunar transient phenomena* (LTP). Table 14–4 on the next page lists nine of the most suspect lunar features and the nature of their events.

LTP have also been reported in Byrgius, Censorinus, Delambre, Euler, Kepler, Linné, Manilius, Menelaus, Messier and Messier A, Pickering, Plinius, Proclus, Pytheas, Römer, Schickard, and other features.

Photoelectric photometry. Amateurs so equipped may wish to make direct photometric measurements of the brightness of select areas of the lunar surface within the umbra and penumbra. Such measurements are

Table 14-4 Suspected LTP Features

Alphonsus	Central peak emissions, dark areas on floor
Aristarchus	Wall bands, "red glow" areas in & nearby
Atlas	Dark areas on floor
Conon	White area on floor
Eratosthenes	Dark areas on floor and walls
Grimaldi	Tone of floor; bright spots on E & NE
Plato	Light spots on floor
Riccioli	Dark areas on floor
Stöfler	Dark areas on floor
Tycho	Glow near central peak

useful for determining the darkness and uniformity of the umbral and penumbral shadows, especially if done at several wavelengths.

Photography. Umbral contacts with the moon's limb are attractive photographic targets for still and video cameras, as are umbral crater contacts and the shadow's progress across larger craters and seas.

Capturing the colors and subtle shading differences of the umbra are worthy goals; try a variety of exposure times with your favorite color films.

You might want to compile a photographic portfolio of favorite lunar features as they appear at frequent intervals throughout the eclipse. Don't neglect including prospective LTP features; you might get lucky and capture an event (a photograph or photo sequence is much more valuable than a subjective LTP visual sighting). Video and CCD imaging are especially encouraged for these projects.

Multiple exposures on a single frame, taken at five or ten minute intervals, can capture an impressive sequence of the partial eclipse's progress. Multiple exposures spaced about an hour apart, taken with a camera driven at a solar (or at least a sidereal) rate, will "freeze" the earth's shadow; the moon's multiple limbs will nearly touch, creating an impressive light–map of the earth's umbral shadow limb and its position in space. Practice on the uneclipsed moon to determine the best exposure interval to prevent overlap of lunar images.

Finally, time–exposures with slow films and small f/–stops result in attractive light–trails of the moon's dwindling, then growing illumination.

For specific exposure recommendations for still photography, see Chapter 17; for video and CCD recommendations, see Chapter 18.

Photograph 14-3 A "moon trail" of the 6 July 1982 total lunar eclipse. A camera, locked into position on a tripod, is set for a time exposure. As the earth rotates, the moon moves across the photograph. And as the eclipse progresses, the light of the moon diminishes. *Photograph taken by Jay Anderson.*

Reference

Graham, Francis G. and Westfall, John E. **Lunar Eclipse Handbook**. East Pittsburgh: Lunar Press, 1990.

Lunar Eclipse Observing – Totality

Introduction. With second contact, the moon completes its journey into the umbra. For as long as the next 104 minutes (but usually a shorter period) the moon will be completely shielded from the sun's direct rays, glowing only with the gray to reddish–orange sunlight the earth's dense atmosphere refracts around the planet's limb.

For all its beauty and the suspense it has generated as to what color and how dark the moon would appear during this particular totality, if we compare it to that of a solar eclipse, second contact is almost anticlimactic.

There are, to be sure, interesting and important projects to pursue during totality, but the partial phases seemed far more hectic. Now, the totality phenomena exhibit themselves for your observation for a considerable length of time while undergoing only the subtlest of changes until third contact's arrival, when the moon's leading edge once more breaks free of its umbral embrace.

Contact timings. As the moon's leading limb approaches the umbra's edge, it is fairly easy to anticipate second contact. Third contact is more challenging, for the moon's trailing limb brightens significantly during the waning minutes of totality. Small apertures, low magnifications and filters are best for enhancing the contrast at the umbra–penumbra boundaries, especially for very bright eclipses.

Attempt timing accuracies to 0.1 min (6 sec) and employ the customary procedure of tape recording voice over WWV or CHU time signals for best results.

General color and uniformity. Carefully characterize the tones and colors of the umbra during totality, while remaining alert to any variations in these characteristics with the passing of time.

To capture the colors of the moon, many observers employ standard paint charts or color wheels illuminated by a white light source. Artistic observers frequently recreate the color by combining watercolors or oils at their palettes until they get the best match. Watercolors seem to recreate the subtle tints best. Photocopies of lunar maps on water–resistant paper stock provide an excellent medium for artists wishing to make frequent watercolor renderings of the changing hues and textures as the eclipse progresses. Once captured, these preliminaries may be transferred to canvas to depict a single moment of totality or, in a sequential fashion resembling a multiple–exposure photograph, to depict the whole of totality.

Monitor various lunar features for visibility within the umbra, especially those which are close to the limiting resolution of your instrument. Employ a variety of color filters to determine which ones best enhance the visibility of specific features.

Eclipse luminosity. The darkness of the moon's surface at mid–totality is governed by the degree to which the earth's atmosphere absorbs the sunlight being refracted around the planet's limb and into the umbra. This, in turn, is determined by how dusty and cloudy the earth's atmosphere is at the time. If there have been major volcanic eruptions, widespread sandstorms, major forest fires or the maximum of a major meteor shower (which may trigger increased cloudiness and rainfall) in the months preceding an eclipse, the moon may be almost invisible during totality, as with eclipses in 1963 and 1964.

A five–point subjective

Photograph 15-1 A near–third contact video image of the 17 August 1989 total lunar eclipse. The VHS video was taken with a Sony black & white camera through a Celestron Comet Catcher. South is towards the left hand corner. *Video taken by John Westfall.*

scale for evaluating the moon's luminosity at mid-totality, proposed by French astronomer André–Louis Danjon (1890–1967), has provided amateurs with an invaluable tool for directly comparing modern lunar eclipses and evaluating written descriptions of historical ones (see Table 15–1).

Table 15-1 The Danjon Luminosity (L) Scale

L = 0	Very dark eclipse. Moon almost invisible, especially at mid–totality.
L = 1	Dark eclipse, gray or brownish in coloration. Details distinguishable only with difficulty.
L = 2	Deep red or rust–colored eclipse. Very dark central shadow, while outer edge of shadow is relatively bright.
L = 3	Brick–red eclipse. Umbral shadow usually has a bright or yellow rim.
L = 4	Very bright copper–red or orange eclipse. Umbral shadow has a bluish, very bright rim.

The strikingly dark eclipses of 30 December 1963 and 24 June 1964 were assigned values of L = 0. More recently, the 30 December 1982 eclipse was rated L = 0.3 while select estimates of the 9 December 1992 eclipse ranged from L = 0 to L = 1.2, illustrating the common practice of using intermediate decimal values with the Danjon scale. Bright eclipses were observed on 16 September 1978 (L = 2.5) and 4 June 1993 (L = 3).

Make your Danjon luminosity estimates with the unaided eye or with low-power binoculars; if you are torn between two of the scale's descriptive criteria, by all means interpolate decimal values. It is recommended that you make one estimate as close to mid–totality as possible, and two others shortly after second contact and just before third contact. These estimates will establish L–values for the outer umbra. Note the times of each estimate and the instrument used, if any.

Apparent magnitude. During totality, the full moon's usual magnitude of –12.7 may drop by a factor from 10,000 to 1,000,000 times; to anywhere from magnitude –2.7 to +3.3 or fainter.

The previous chapter dealt extensively with apparent magnitude estimates of the partial phases; those techniques should be carefully reviewed and applied during totality as well. Use the method and comparison stars best suited to the unique conditions of each totality.

Lunar transient phenomena (LTP). Chapter 14 also dealt with this important area. Monitor the lunar features identified in Table 14–4 during totality as well as during the partial phases. You'll find you not only have more time for the task, but the umbral dimming and color can make these subtle changes more apparent.

Stellar total and grazing occultations. During totality, the moon's limbs are dark enough to make the disappearances and reappearances of faint stars encountered by the moon's inexorable eastward drift through the heavens readily visible.

Occultations are predicted well in advance and widely disseminated through astronomical publications, or are available for your viewing site through the International Occultation Timing Association (I.O.T.A.). A nominal fee is charged for non-members.

Timings require access to WWV or CHU radio time signals, a tape recorder, and a telescope adequate to the task, and timings should be made to the nearest 0.1 second with an accuracy as close to 0.2 second as possible. You also need to obtain the precise latitude, longitude and elevation of your observing site from a topographic map if your observations are to have any value.

Stars will disappear along the moon's western limb (celestial east) and reappear at the eastern limb (celestial west) for total occultations, and will be close to the polar regions for *grazing occultations*. Grazes present greater challenges, for you generally have to travel some distance to get the star and the moon in proper alignment and there may be multiple disappearances and reappearances in rapid succession as the moon's mountains and valleys pass your line of sight.

On rare occasions, planets, asteroids, comets and deep-sky objects may be occulted during totality; these can be especially exciting and observationally important events.

I.O.T.A. is especially interested in receiving timings of grazing and total occultations made during lunar eclipses.

Photograph 15-2 The totally–eclipsed moon in Scorpius, 25 May 1975. This 4 minute exposure was taken on GAF 400 through a 55 mm lens at f/1.8. North is to the top. *Photograph taken by John Westfall.*

Meteors. Meteor observations take two forms. You might want to watch for large meteor impacts on the moon's surface (of course, you can do so for much longer periods on the moon's earthlit surface between full moons). Or, you might prefer observing sporadic or shower meteors which would otherwise be "washed out" by the full moon's light were no eclipse in progress.

Lunar meteor impacts are best watched for by remaining alert to their possibility while conducting other observations, most notably high-power monitoring of LTP and other features with large-aperture instruments. Large impacts are considered exceedingly rare and, even if one occurs, likely to be misinterpreted as an LTP of lunar origin.

Observers with modest equipment and an interest in meteor observing may want to stretch out in a lawn chair during totality and monitor earthly meteor activity. An hourly count might be made, or more extensive data on the times (to the nearest minute), magnitudes, durations, train characteristics and paths of observed meteors may be recorded. These data may be especially important if the eclipse falls near the maximum of a major meteor shower.

Others may want to use binoculars or telescopes to watch for fainter telescopic meteors. Select a familiar star field having stars of wide-ranging magnitudes to simplify estimates of meteor brightness.

Meteor observations should be forwarded to one or more of the following: the American Meteor Society, the Meteor Section of the Association of Lunar and Planetary Observers, the International Meteor Organization, and the newsletter **Meteor News**.

Variable stars. The fainter novae, short-period, and irregular variable stars might be observed and their magnitudes estimated during totality if the cycle of the full moon is likely to make them otherwise unobservable for a week or so. Report observations to the American Association of Variable Star Observers.

Artificial satellites. In addition to the brighter artificial satellites, which are easily seen even against the fully-illuminated full moon, totality presents an opportunity to observe some of the fainter and higher ones which may be having favorable passes at the time. Most satellite enthusiasts have computer software which generates predictions for their sites.

Photoelectric photometry. As with the partial phases, photometric measurements of the umbra should continue during totality. Singling out specific lunar features for photometric monitoring, especially those on the LTP list, can yield very valuable confirmation of any changes detected by visual observers. You may also want to map the brightness variations within the umbra. The International Amateur-Professional Photoelectric Photometry organization can provide guidance.

Photography. In addition to the multiple and trail exposures recommended in Chapter 14, telescopic photography of the moon at all available magnifications should be attempted with still, video and CCD systems.

Amateurs with more than one camera may want to use a variety of different color and black-and-white emulsions for still photography; others will have ample time to change film types during totality.

Particularly spectacular occultations might also be photographed, either with a sequence of exposures or by guiding on the moon and letting the star trail until it disappears, or letting it trail after it reappears.

Short time-exposures, around 5 minutes, with wide-field and standard lenses, should capture bright meteor activity without serious film fogging.

See Chapter 17 for still-photography and Chapter 18 for video and CCD guidelines.

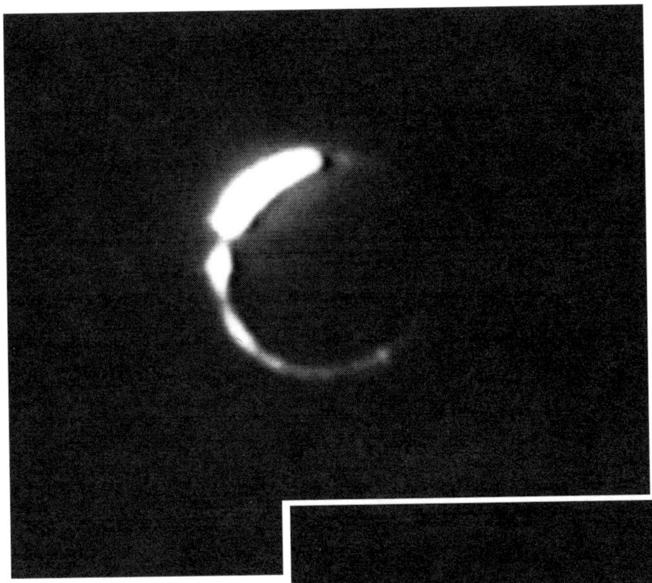

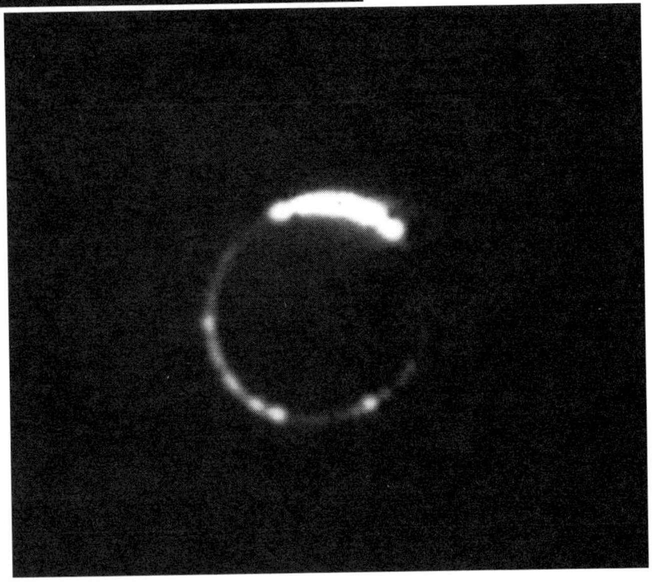

Photograph 15-3a, b
Unusual views of a total lunar eclipse: a total solar eclipse as seen from the moon! Surveyor 3 images of the 24 April 1967 eclipse. The above image was taken at 11:24 UT; the right image at 12:01 UT. A total lunar eclipse was visible across East Asia, Australia and the Pacific. *Photographs courtesy of the National Aeronautics and Space Administration.*

Eclipses of Other Types

Introduction. There are many types of eclipses that can be observed by amateurs which do not involve the sun, moon and earth alone. These can be loosely classified as planetary transits, eclipses of planetary satellites, occultations, eclipsing binary stars and spacecraft events.

Planetary transits. The planets Mercury and Venus revolve about the sun in orbits inside the orbit of the earth. If one of these planets is at the ascending or descending node of its path about the same time as it is in *inferior conjunction* with the sun, an observer on the earth will see the planet *transit* the sun.

The first prediction of a transit of Mercury was made by the astronomer Johannes Kepler for the 7 November 1631 transit, which was observed from Paris by Pierre Gassendi. The first transit to be accurately timed was that of 1677.

Transits of Mercury occur at ascending node in early May, at intervals of 13 or 46 years, when the planet is near aphelion; descending node transits, which are more frequent, occur in early November, at intervals of 7, 13 or 46 years, near perihelion (see Table 16–1).

A May transit of Mercury, crossing the center of the sun's disk, may last nearly nine hours, while a November transit barely grazing the sun's limb may be very brief, as with the 55–minute 1999 transit.

The first prediction of a transit of Venus came in 1627, again by Kepler, for the transit of 7 December 1631. Gassendi failed to observe it, however, as it ended just before dawn in Paris. The first to be observed was that of 4 December 1639 by English amateurs Jeremiah Horrocks and William Crabtree. The 1761 transit was the first observed in North America, while the transit of 1769 was a justification for Captain James Cook's three–year voyage aboard HMS *Endeavour*.

Transits of Venus are less frequent (see Table 16–2),

Table 16-2 Transits of Venus

7 Dec 1631	6 Jun 1761	9 Dec 1874	8 Jun 2004	11 Dec 2117
4 Dec 1639	3 Jun 1769	6 Dec 1882	6 Jun 2012	8 Dec 2125

occuring in pairs eight years apart at alternating intervals of 113 and 130 years. Ascending node transits occur in early June and descending node transits in early December.

If Mars is colonized by 10 November 2084, explorers of the red planet will have an opportunity to observe a transit of the earth and moon across the sun. Arthur C. Clarke's 1971 science–fiction short–story *Transit of Earth* featured the prior one of 11 May 1984; the following earth transit will be on 15 November 2163.

Observations of planetary transits must be conducted with the same attention to eye protection as those of solar eclipses; full–aperture front–mounted mylar or glass filters are recommended for telescopes and binoculars. Transits of Venus are visible with the unaided eye, but Mercury's smaller angular diameter requires a telescope to see.

The four contacts, when the transiting planet's leading and trailing limbs first move onto and off of the sun's disk, should be timed as accurately as possible. You might want to time contacts with major sunspots or sunspot groups as well.

Second and third contact timings are complicated by the "black drop" effect where the planet's limb seems to delay its separation from or hasten its reunion with the sun's limb. This optical effect, similar to a drop of water clinging to an object before breaking away, is attributed to irradiation and refraction of sunlight in the earth's atmosphere.

Lomonosov's phenomenon is a crescent–shaped band of light resembling a blister on

Table 16-1 Transits of Mercury (UT)*

7 Nov 1631	11 Nov 1736	9 Nov 1802	10 May 1891	9 May 1970
9 Nov 1644	2 May 1740	12 Nov 1815	10 Nov 1894	10 Nov 1973
3 Nov 1651	5 Nov 1743	5 Nov 1822	14 Nov 1907	13 Nov 1986
3 May 1661	6 May 1763	5 May 1832	7 Nov 1914	6 Nov 1993
4 Nov 1664	7 Nov 1756	7 Nov 1835	8 May 1924	15 Nov 1999
7 Nov 1677	9 Nov 1769	8 May 1845	10 Nov 1927	7 May 2003
10 Nov 1690	2 Nov 1776	9 Nov 1848	11 May 1937	8 Nov 2006
3 Nov 1697	12 Nov 1782	12 Nov 1861	11 May 1940	9 May 2016
5 May 1707	4 May 1786	5 Nov 1868	14 Nov 1953	11 Nov 2019
6 Nov 1710	5 May 1789	6 May 1878	6 May 1957	13 Nov 2032
9 Nov 1723	7 May 1799	8 Nov 1881	7 Nov 1960	7 Nov 2039

*Dates are determined by the Universal Time (UT) of mid–transit; actual calendar dates may vary by a day for specific observing sites on the earth.

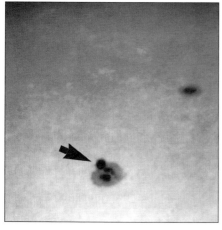 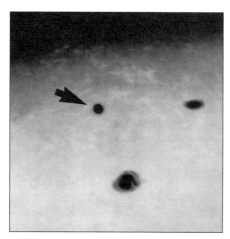 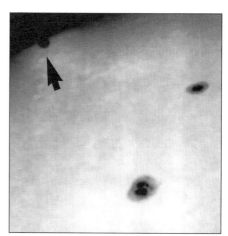

Photograph 16-1a–c The 9 May 1970 transit of Mercury. Note the location of Mercury near the sunspot group; left and at third contact, right. *Photographs taken by J. S. Korintus,* ©*The Royal Astronomical Society.*

the sun's limb which Venus appears to push ahead of it at third contact. The probable cause is irradiation where two areas of very different brightness are observed in close proximity to one another under poor conditions.

The thick atmosphere of Venus is believed to be the cause of a delicate luminous halo or ring of light, called an aureole, which surrounds the planet just as it is moving onto or off of the sun's disk. Mercury shows no such halo, and Mariner 10 spacecraft measurements in 1974 and 1975 and subsequent earth–based studies confirm that Mercury has only a very tenuous atmosphere at best.

Photographers should capture transits with exposures and systems which work well for sunspot imaging.

Eclipses of planetary satellites. The four Galilean satellites of the planet Jupiter, Io, Europa, Ganymede and Callisto, and some of the brighter satellites of Saturn, Tethys, Dione, and Rhea, undergo eclipse phenomena which can be viewed through moderate–sized telescopes.

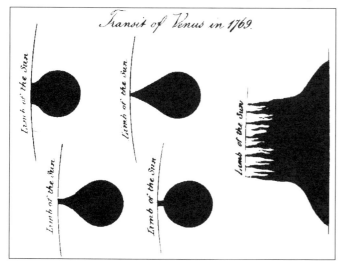

Figure 16-2 Observation drawings by S. Dunn of the black drop effect, 1769 transit of Venus. *Drawings* ©*The Royal Astronomical Society.*

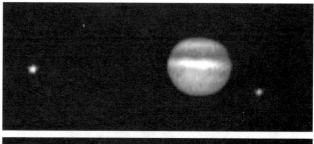

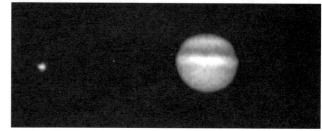

Photograph 16-3a–d The satellite Europa, right of Jupiter, can be seen passing into Jupiter's shadow, 7 April 1994. These 0.05 second CCD images with no filter were taken once every 60 seconds with a LYNXX CCD through a 28 cm Schmidt–Cassegrain at f/10. Note Io to the left of Jupiter. North is up. *CCD image series taken by John Westfall.*

Amateurs are most familiar with the eclipses of Jupiter's moons from the monthly predictions appearing in popular astronomical periodicals. The moons may be observed passing into and out of Jupiter's shadow, being occulted by the planet itself, and transiting, along with their more prominent shadows, across the giant planet's cloud belts. The satellites can also, on very rare occasions, eclipse and occult one another.

Eclipses of Saturn's satellites by the ringed planet's shadow are less frequent and their fainter magnitudes make their events more difficult to observe. **The A.L.P.O. Solar System Ephemeris**, an annual publication of the Association of Lunar and Planetary Observers, is one source for predicted times.

Timings of eclipse events are the major observations amateurs can make here. Photoelectric photometry can be an invaluable approach to measuring the magnitude changes the satellites undergo during eclipses; visual magnitude estimates, however, are very difficult to make unless there are suitable comparison stars nearby.

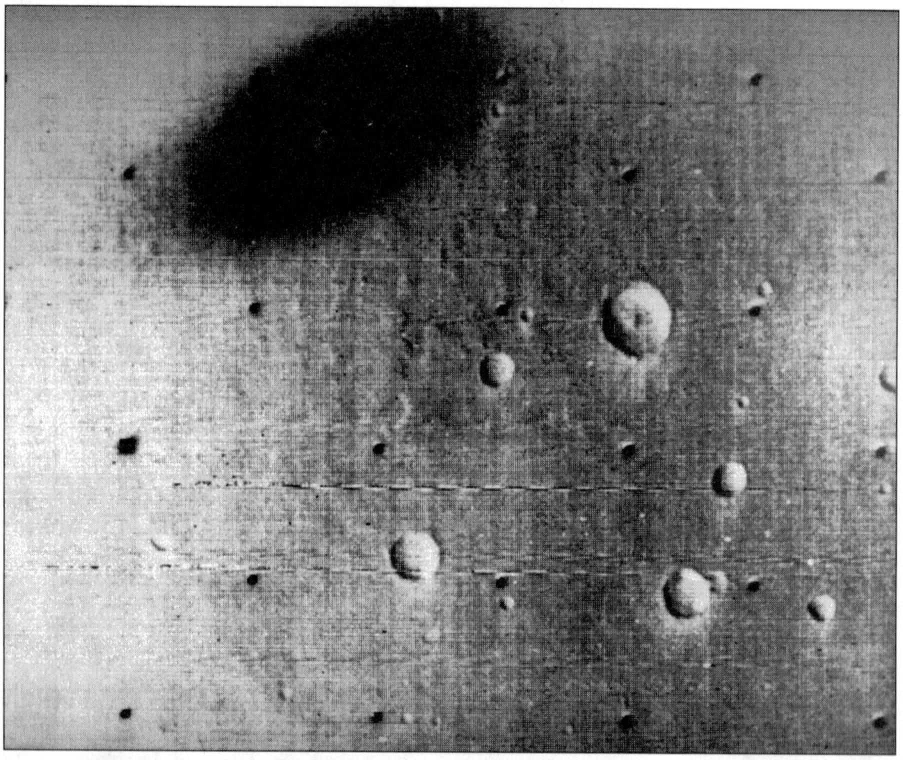

Photograph 16-4 The elongated shadow of the Martian satellite Phobos, 50 kilometers by 11 kilometers, can be seen easily in this February 1972 Mariner 9 spacecraft image. *Photograph courtesy of the National Aeronautics and Space Administration.*

Occultations. Stellar and planetary grazing and total occultations by the eclipsed moon are discussed in Chapter 15. Interested amateurs are urged to observe and time similar events when the moon is out of eclipse as well. Disappearances and reappearances of all but the brightest stars should be timed against the moon's unilluminated limb. For disappearances, this is between the waxing crescent and nearly-full moon in the evening sky; for reappearances, opportunities begin just after full moon and continue through the waning crescent phase in the pre-dawn sky.

Observers can make significant discoveries by being alert to unexpected step-like dimmings during occultations, indicating the presence of a previously unknown companion star.

Planets will occasionally pass in front of stars bright enough not to be lost in their glare. When this happens, stars may disappear abruptly if the planet has little atmosphere, or fade gradually as it is obscured by increasingly thicker atmospheric layers. The planet Saturn's rings sometimes occult stars as well, providing outstanding opportunities to measure the particle densities of the various rings and their gaps. Photometric observations are especially desired for planet, planetary satellite and ring occultations, and photographic, video, and CCD imagery can capture these impressive events.

Occultations of stars by dimmer asteroids can be observed with any instrument with a magnitude reach capable of detecting the star. Being able to see the asteroid is not a requirement, although it increases the viewing experience. Asteroid occultations, like solar eclipses, rarely come to the observer. The shadow path of the asteroid across the earth can be quite narrow and not very accurately known until days before the occultation. The International Occultation Timing Association provides members with predictions and a telephone hotline for path updates.

Observations of asteroid occultations are conducted by traveling to the path of visibility, setting up, locating and tracking the target star early, then tape-recording the observed disappearance and reappearance against broadcast time signals (the interval between the two events will usually be less than ten seconds for all but the largest asteroids).

Observers should be alert to possible very brief wink-outs of the star shortly before or after the main event. The discovery by the Galileo spacecraft in 1994 of a moon orbiting the asteroid Ida, similar indications by radar probes of earth-passing asteroids, and the linear arrays of large craters on many planets and their moons, lends credence to reports of possible satellites of asteroids by observers of past asteroid occultations.

Occultations of stars by the coma and tails of comets have been observed frequently. Stars have been observed to dim as the coma passes over them. The fluorescent gas (ion) tail of a comet is so diffuse that little dimming is noticed except against the brighter sections; the dust tail,

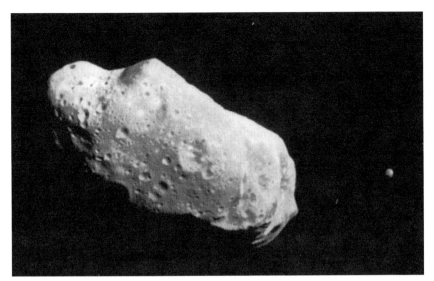

Photograph 16-5 The asteroid Ida and its moon Dactyl. Such asteroid systems present potentially–unusual occultations when an asteroid passes in front of a star. *Photograph courtesy of the National Aeronautics and Space Administration.*

composed of solid particles from the nucleus, is more likely to cause dimming. The very small nuclei of most comets, combined with the brightness of their surrounding comas, all but preclude true occultations, however.

Eclipsing binary stars. Many stars are binaries, meaning two stars gravitationally linked and orbiting about a common center of gravity. If the plane of their common orbit is aligned with the earth, the two stars may be observed to partially or totally eclipse one another. From the earth, only a single star is seen, of course, but the presence of its companion can be determined by the variations in brightness observed.

When the two stars are out of eclipse their combined light will be at maximum, but twice each orbit, when one passes in front of the other, the combined light will dim. By plotting these dimmings (called "minima") with respect to time, the orbital period of the pair can be determined.

If the companions differ significantly in brightness, the two minima will be different as well, a shallow fading when the brighter twin covers the dimmer, and a deeper fading when the brighter star is being eclipsed.

The American Association of Variable Star Observers is the umbrella organization for observers interested in becoming involved with this and other types of variable star observing.

Spacecraft events. Artificial satellites are frequently observed occulting stars. They may also be eclipsed by the earth's shadow, occult one another, and transit the face of the moon and planets.

Accurately–timed observations of satellite occultations and transits can define the orbits of satellites and space debris which do not have active radio transmitters. Observations of satellite eclipses by the earth's shadow can define the shadow's location and the degree of atmospheric refraction at times other than lunar eclipses.

Satellites brighter than second magnitude will remain visible throughout a transit of the moon's illuminated disk, providing a striking illustration of just how dark the moon's surface really is. Dimmer satellites will be visible over the darker lunar maria ("seas"), but will be obscured by the brighter highlands and rays.

It is easier to observe satellites entering the earth's shadow than leaving it, but many home computer software programs can provide precise look–angles or celestial coordinates making even reappearances simple. Satellites will not disappear or reappear abruptly, but will undergo gradual fading and brightening instead.

Still and video photography of the brighter satellites is easily accomplished. Time–exposures with sturdily–mounted still cameras will capture impressive trails of satellites entering or leaving the shadow against background starfields. Video cameras should be operated manually for exposure and focus.

A fortunate photographer capturing several satellites at slightly different altitudes on the same frame, within a few minutes of one another, will have an impressive "map" of the earth's shadow cone against the sky.

Fainter (and/or much higher) satellites may be captured through telescopic lenses and telescopes, or with CCD's. Geosynchronous satellites are within reach of intermediate–to–large amateur instruments; once located, observations and telescopic photographs will show the unmoving satellite in the center of the field while the earth's rotation causes stars to sweep past. Stars brighter than the satellite may briefly wink out in occultation.

Satellite prediction software is accurate enough to capture a passage within the 0.5–degree field of view of low–power telescope eyepieces with surprising regularity, but your software can be no better than the currency of its satellite element file; download updated elements from computer bulletin boards at least monthly.

References

Clarke, Arthur C. *"Transit of Earth"* (originally published in 1971 in **Playboy**), in **The Ascent of Wonder: The Evolution of Hard SF** (David G. Hartwell and Kathryn Cramer, editors). New York: Tor/Tom Doherty Associates, Inc., 1994.

King–Hele, Desmond. **Observing Earth Satellites**. New York: Van Nostrand Reinhold Company, 1983.

Meeus, Jean. **Transits**. Richmond, VA: Willmann–Bell, Inc., 1989.

Sweetsir, Richard. *"Letter"* (observation of the Echo I satellite transiting the moon). **Sky & Telescope** (May, 1962).

Figure 16-6 A once–in–a–lifetime eclipse: a solar eclipse created by the Apollo spacecraft as seen by cosmonauts Aleksey Leonov and Valeriy Kubasov aboard Soyuz on 19 July 1975 during the Apollo–Soyuz Test Project. *Illustration by George Murphy.*

17

Eclipse Photography

Introduction. One of the joys of eclipse observing is the opportunity to relive the eclipse later while sharing the excitement of your experiences with others. Eclipse photography allows you to accomplish this goal. Chapter 16 overviewed opportunities to photograph other eclipse phenomena; this chapter addresses solar and lunar eclipse photography exclusively.

Photography may be on a simple scale for the novice, or involve exotic equipment and techniques yielding fabulous results for the expert. The simplest approaches are unlikely to produce magazine–cover quality photographs but will give you something to "show the folks at home." One word of advice, however; remember to **look** at the eclipse too. Your only recollection of totality should not be an image you glimpsed through the restricted viewfinder of your camera or the finder scope of your camera–laden telescope!

Film. Your choice of film depends upon the type of eclipse, the equipment you plan to use and the results you desire. Advances in photographic emulsions give today's astronomer a variety of films from which to choose. You need to determine if you want black & white (don't discount it yet), color prints or color slides; all offer unique qualities that warrant consideration.

New emulsions appear on the market all the time. Look for them and the advantages they might offer. Technical publications available from film manufacturers and articles in photography magazines will assist your review of current and new emulsions.

• **Black & white emulsions:** Off–the–shelf black & white films offer a variety of film speeds, denoted by the ISO (previously ASA) of the film. Kodak's Plus–X film gives a moderate speed of ISO 125, whereas Kodak Tri–X is much faster at ISO 400. But don't just jump on the faster–is–better bandwagon. The fine–grain feature of Plus–X will produce much finer details or resolution than Tri–X. Film resolution is a more important factor for a lunar eclipse than a partial or annular solar eclipse.

Other black & white emulsions currently available include Agfapan 400 Professional, Ilford HP5, Kodak T Max, and Kodak Technical Pan 2415 (preferred by planetary photographers for its very fine grain).

• **Color print films:** Films identified by the suffix "color" (e.g., Kodacolor and Fujicolor) yield color negatives and prints while providing the astrophotographer with the widest possible exposure latitude. They are available in a selection of speeds and their popularity for everyday use has made custom one–hour processing and enlarging available in most communities. Kodak Royal Gold films are widely acclaimed for their high quality and color balance. Also consider using professional film; this is usually stored refrigerated at your local camera store and will cost a bit more than off–the–shelf film.

• **Color slide films:** Transparency films, identified by the suffix "chrome" (e.g., Kodak's Kodachrome and Ektachrome, Fuji's Fujichrome, and 3M's ScotchChrome 800/3200P), produce color slides and are a good choice if your plans include sharing your experiences with large groups. These films are also available in a range of speeds, in both off–the–shelf and professional versions. Most photo processors can produce excellent prints and enlargements from your transparencies quickly and inexpensively.

Although slower–speed films produce better images, a number of eclipse chasers, especially first–timers, opt for faster films to increase the odds of taking an acceptable photograph. Faster speeds allow for shorter exposure times, important if the system is "slow" or vibrations are a potential problem. Color balance of some slide films is an issue you might want to research further.

Photographic equipment. A simple 35 mm camera with a standard 50 mm focal length lens will form an image of the eclipsed sun or moon on your film having a diameter of less than 0.5 mm. Enlarging images of that size will generally produce unsatisfactory results unless you are making multiple exposures at frequent intervals on a single frame. A telephoto lens of 200 mm will provide a reasonable image, but the camera–lens system should be tripod–mounted to reduce vibrations. This is essential for total lunar eclipses which require longer exposures even with fast films.

• **Cameras:** The choice of most eclipse chasers is the 35 mm single–lens reflex camera, or SLR. The 35 mm refers to the width of the film, whereas a SLR allows the photographer to interchange lenses and view the scene directly through the lens. The 35 mm SLR provides a system adaptable to almost any requirement, employs reasonably sized film, and is available in a variety of systems to accommodate most budgets.

Avoid SLR's having automatic focus, exposure and aperture features. Unless they have a manual override capability, they will present problems for eclipse photography.

There are a variety of camera accessories available to

Photograph 17-1 The partial phase of the 10 May 1994 annular eclipse. This exposure is through a standard 50 mm lens on Ektachrome ISO 100. *Photograph taken by Mike Martinez.*

Photograph 17-2 This exposure of the same eclipse as **Photograph 17-1** except taken through a 200 mm lens with a 3X telextender, thus producing a more-desirable photograph. *Photograph taken by Mike Martinez.*

eclipse observers. Data backs imprint specifics (date, time, etc.) on the film. An automatic advance or auto winder can be an excellent accessory to reduce vibration and to position the film for the next exposure rapidly. A cable release will reduce vibrations even further. Some models have a manual mirror-raising feature to further reduce vibrations from mirror slap, an advantage for exposures through telescopes.

Large format cameras will produce even better results. Since their negative size is significantly larger than that of their 35 mm counterparts, enlargements tend to be more spectacular. However, cost can be a major factor with such systems.

•**Longer focal length systems:** We recommend that longer focal length lenses or a telescope be employed for eclipse photography. A 2000 mm focal length, that of most 200 mm (8 in) diameter commercial Schmidt-Cassegrain systems, will produce a solar or lunar image that almost fills a 35 mm frame. Refer to Figure 17-6a, b to choose lens.

The ideal method of eclipse photography is directly through the telescope. Even a simple telescope offers a longer focal length and (usually) greater light gathering power than most telephoto lenses. Many telescopes also offer the advantage of a sturdy equatorial mount and variable-speed clock drive, useful for tracking the eclipsed sun or moon and essential for long exposures during a dark total lunar eclipse. If you're traveling to the Southern Hemisphere, make certain your clock drive is reversable.

The easiest method of photography through the telescope is *prime focus* (a.k.a. direct objective). With this method, you remove the telescope's eyepiece and the camera's lens; the telescope's objective lens or mirror serves as the camera's lens. A number of camera adapters are commercially available that attach directly to the camera's body and have the same outside diameter as the telescope's eyepiece, allowing the camera adapter to be inserted directly into the eyepiece holder as you would an eyepiece.

Other adapters are available that allow you to retain the eyepiece (called positive projection) or to use a barlow lens or photographic telextenders (negative projection) to increase magnification. Such systems introduce longer exposures but can be advantageous if you want to focus on a spectacular solar prominence, sunspot group or lunar feature. The optical quality of photographic telextenders are occasionally poor so you might consider other alternatives. Use caution when selecting your eyepiece also; excessive magnification can produce fainter images and accentuate vibrations.

Finally, if you are using a camera with a fixed lens, employ an *afocal system* in which the lens is aligned with an eyepiece to produce an image. Such alignment is tricky, requires home-made light baffles to keep out external reflections, and introduces even longer exposures than the positive or negative projection systems.

There are many excellent books on the subject of astrophotography that cover film, cameras and photographic systems as well as other options for the eclipse chaser.

Focusing. The eye is not always a reliable tool for obtaining razor-sharp focusing through a telescope, especially when viewing a dark total lunar eclipse. Focus is not maintained throughout a long photographic session either; temperature changes, the weight of your camera equipment on a rack-and-pinion focuser, even flexure of light-weight telescope tubes can alter the focus. Re-focus frequently during an eclipse *except* during the total phase of a solar eclipse. Helical focusers are preferable to a rack-and-pinion type; you can obtain finer corrections with their circular screw thread arrangement and they are not subject to slipping under the weight of a camera. If your camera has interchangeable focusing screens, a fine

Photograph 17-3 Not the type of multiple exposure one wants! Mounting vibrations ruined this exposure of the 10 May 1994 annular eclipse.

ground glass screen, used in conjunction with a focus magnifier, works well; otherwise, a high-power magnifier with your existing screen may be your best option. A knife-edge focuser does an excellent job, but this requires focusing on a star, then moving the telescope back to the eclipse for your exposures; adequate for lunar eclipses, but impractical for solar eclipses. Finally, keep your film tight in your camera (turn the rewind lever until you feel it resist); this will take up any slack and flatten the film against the film plane.

Photographing a solar eclipse. Solar eclipses have two distinctly different stages to photograph: the partial (including annular) phase and totality itself require separate guidelines.

•**The partial and/or annular phases:** Any photography of the sun's brilliant disk requires a specialized filter to protect your eyes. You **must** follow all safety rules; a camera will **not** protect the eye from the sun's harmful radiation. Some **photographic** filters recommended for solar photography shorten the required exposure, provide pleasing images, *but do not protect the eyes*. If using these, it is essential that you sight through a safe **visual** filter placed in front of the camera lens and/or viewfinder while centering and focusing the sun; then replace the visual filter with the photographic one for the exposure. Use of a safe visual filter is also essential when composing prior to photographing the sun through clouds or at sunrise or sunset when the sun is significantly dimmer.

The color of the sun in your photographs can range from orange or yellow to a blue or blue-green tint, depending upon the type of filter used; the film's color balance is usually not an issue during the partial or mid-annular phases.

Determine in advance how often you want to photograph the partial phases; many eclipse chasers will document them every five or ten minutes. A series of photographs of the moon eclipsing a sunspot group, taken at higher magnification using eyepiece or negative projection, would enhance the series. Plan your photography in advance, and be familiar with your system; you do not want to run out of film and be fumbling around trying to reload just before mid-eclipse! Many observers use bulk-film backs for their cameras, an option to consider if you have only one camera (when establishing your photographic eclipse plans, determine where you can have your bulk film processed so you won't experience any problems upon your return home!).

Focus is also an issue. Even with a solar filter, heat builds up in a telescope tracking the sun, and this can change the focus of the instrument. Continue to check the focus as the eclipse progresses.

An option to consider is photographing projected images of the eclipsed sun. Pinhole images produced by tree foliage, as well as telescopic projections onto screens (or a fellow observer's white T-shirt) provide not only a safe way to photograph and view the partial phases but an interesting photographic alternative.

•**Typical exposures:** The length of your exposure depends upon the speed of the primary objective, the speed of the film, the type of solar filter in use, the type of photographic set-up (afocal, direct objective or projection), and the percentage of the sun eclipsed.

By photographing the uneclipsed sun through the system you plan to use at an eclipse, you can practice photography of the partial or annular eclipse. Exposures between 1/125 and 1/1000 second at f/8 or f/11 (for ISO 200 film) are good departure points for evaluating your system.

•**Totality:** Totality is an exciting and wonderful time; it is advantageous that you carefully plan your photography well prior to the eclipse. Don't try to do it all; spend a little time enjoying the eclipse, and **PRACTICE** in advance!

We recommend that you use a minimum 400 mm focal length instrument that is at least tripod-mounted. It is best to use a portable battery-powered clock-driven equatorial mount if possible. Check in advance (*but don't count on!*) on the availability of 110-120 volt 60 hertz alternating current (AC) if your planned observing site is outside the United States; you may find it necessary to invest in a power adapter for international current standards or carry along a battery system.

•**The diamond ring and Baily's beads:** Exposures of 1/125 or 1/250 second at f/8 will capture the diamond ring and Baily's beads on ISO 200 film (see Figure 17-6b). These exposures will be much shorter than those you will employ during totality. Be prepared to quickly and efficiently remove the solar filter (being cautious not to look at the last remaining seconds of the partial phases!) after adjusting your f/stop and exposure settings for the much fainter phenomena of totality.

Internal reflections have detracted from many otherwise-excellent diamond ring photographs, especially if the lens used contains a large number of elements. Evaluate your system by photographing a bright light bulb, streetlight, or a tiny part of the unfiltered sun through a simple "artificial diamond ring" device made from a cardboard tube fitted over your lens and an end cap with a small hole punched in it (observing all vision safety precautions, of course). Otherwise, you risk multiple diamond ring ghost images on each frame.

Photograph 17-4 Too many diamond rings! Ghost images were created by internal lens reflections on this exposure of the 20 June 1974 total solar eclipse.

• **The corona, prominences and chromosphere:** An ideal exposure for the brilliant white outer corona and streamers will capture little of the inner corona's delicate nature and will also lessen your chances with the spectacular solar prominences. By progressively shortening each subsequent exposure, however, you will capture the subtle structural variations within the corona and do justice to the chromosphere and prominences as well. Figure17–6a, b offers some recommended exposure times for films of ISO 64 and 200. For other ISO film speeds, recall that each time you double the ISO you decrease the f/stop by one stop.

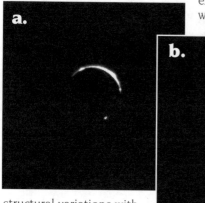

Photography of other solar eclipse phenomena. Don't overlook the opportunity to photograph other eclipse phenomena, as well as other ways to document the eclipse.

• **Multiple exposures:** A spectacular way to record an eclipse is through a multiple exposure. Your camera must be capable of taking more than one exposure per frame and must be firmly fixed to avoid any motion. A cable release will further minimize vibration.

You will need to determine the field of view of your lens. Since the sun's motion across the sky is 15 degrees per hour, you can then calculate the length of time it will take the sun to cross your lens' field of view. A 50 mm camera lens has a field of view of approximately 45 degrees on the diagonal; it will take the sun about three hours to cross this field. Determine in advance how often you want to expose your film to the progressing eclipse; every five to ten minutes works well. You'll want to orient your camera so that the earth's rotation carries the sun through the center of your field and along the field's widest axis. You will probably want mid–eclipse to occur with the sun near the center of your field; estimating this position in the sky and centering your camera on this point will help you determine when to make your first exposure on the frame and what exposure interval to use. For a total eclipse, don't forget to remove the sun filter for the exposure taken during totality!

• **Shadow bands:** The equipment required for shadow band photography includes a camera with a normal focal length lens of around 50 mm, fast film of at least ISO 400, a white screen made of a material like plywood that will not blow away, a meter stick placed across the white screen, and much luck!

Exposures should be 1/250 second or less; try to bracket exposure and/or aperture, if possible, while taking photographs as quickly as possible. An auto winder and a data back, calibrated with broadcast time signals prior to totality in order to record the time (in seconds, if possible) on each photograph, will be useful. See Chapter 7 for additional details.

• **Atmospheric and sunrise–sunset effects:** Spectacular photographs of horizon colors, or the sunrise–sunset effect, are inviting targets prior, during and just after totality. A wide angle or fisheye lens provides superb images; a series of photographs will show the colors as they move around the

Photograph 17-5a–g
This series of photographs of the 20 June 1974 total solar eclipse demonstrates how changes in exposure produce varying affects.
Photograph 17-5a was exposed at 1/125 second, **17-5b**: 1/60 second, **17-5c**: 1/30 second, **17-5d**: 1/15 second, **17-5e**: 1/8 second, **17-5f**: 1/2 second and **17-5g**: 1 second exposure. *Photographs taken by Carter Roberts.*

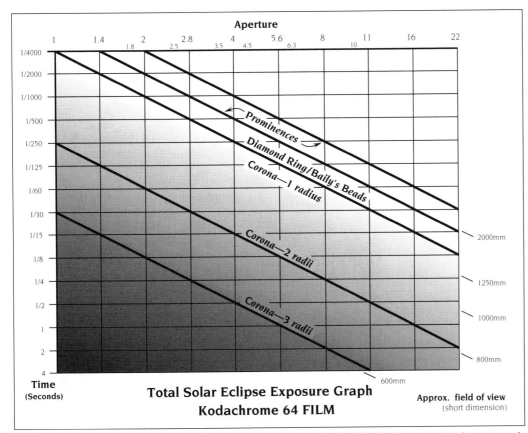

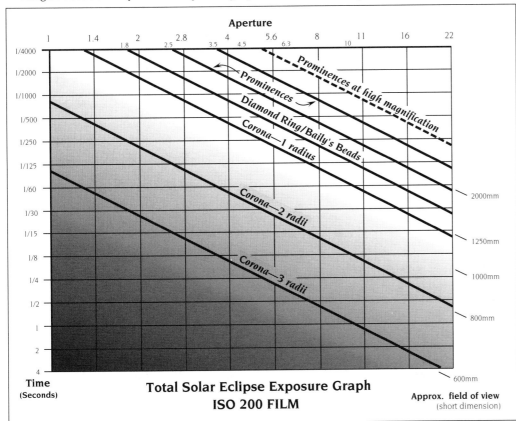

Figure 17-6a, b: Total solar eclipse exposure graph for Kodachrome 64 above, and ISO 200 films below, showing exposure length versus aperture. Note exposure length versus eclipse detail photographed. *Illustrations by Carter Roberts.*

horizon. The camera should be tripod–mounted; as the moon's shadow ap-proaches your site the exposures will lengthen. You can also obtain a fish-eye effect by photograping a hemispherical hubcap through a standard lens on a tripod–mounted camera.

An impressive series of photographs showing the darkness of totality results from not changing the exposure or aperture; the initial and final photographs will be a little overexposed, while those during totality will be a little underexposed.

One of the few positive aspects of a cloudy eclipse day is photographing the impressive motion of the moon's shadow across the sky. A bright cloud deck provides contrast, enhancing the shadow's appearance and making photography easy! Many observers of the 7 March 1970 and 26 February 1979 total solar eclipses were treated to this consolation prize for their efforts. A wide angle or fisheye lens, or the hubcap method, yields the best results.

•**Planets, stars and other objects near the totally eclipsed sun:** A standard lens and a long exposure will record not only the totally eclipsed sun but bright stars, planets and maybe even a comet! The camera should be tripod–mounted and you should use a cable release; exposures of 10 to 20 seconds should provide excellent images without excessive sky fogging from the sun's bright corona. Practice with simple night-time photography of stars and planets with your lens system to determine the maximum exposure possible before the earth's rotation makes trailing of

63

star images evident (probably around 10–20 seconds).

Photographing a lunar eclipse. As with a solar eclipse, you will find distinctly different aspects to photograph; the unpredictability of the darkness of the umbral shadow can make lunar eclipses a challenge as well.

•**The partial phases:** Photography of the early penumbral stages of the eclipse is difficult; we suggest you use film with the widest latitude. The problem is trying to capture the faint, hard–to–detect penumbra against the brilliant uneclipsed lunar surface. As the eclipse proceeds, you will have better luck isolating the penumbral shading photographically.

Once the umbra begins to cross the moon, another challenge presents itself. The intensity difference between the penumbra and umbra is a factor of about 10,000; your best opportunity is just before the umbral shadow covers the moon and a small percentage of penumbral shadow remains.

The difficulty in photographing the umbra's progression lies in the fact that there is quite a brightness contrast between the regions of the moon within and outside the umbra. You can expose for the darkness of the umbra, the brightness outside the umbra's shadow, or both. If the umbral shadow is colorful, a color print film, with its wider exposure latitude, will provide the best results. However, Kodak Technical Pan 2415, a black–and–white film, has been successfully used for penumbral and partial eclipse photography.

•**Totality:** The brightness of the totally eclipsed moon can vary greatly, and the intensity of the earth's shadow itself may not be uniform. The 6 July 1982 eclipse varied from a red hue on one limb to almost white on the opposite limb!

A very dark eclipse may require long exposures of up to several minutes (depending upon the focal ratio of your system and the film speed). A clock–driven system that tracks to the rate of the moon's motion is necessary for exposures of such durations.

We recommend photographic systems with at least a 500 mm focal length. Systems in the range of 1000 to 2000 mm will yield a highly resolved view of various lunar features.

Photography of other lunar eclipse phenomena. Total lunar eclipses offer opportunities for additional photography to supplement your record of the eclipse itself.

•**Planets and stars during totality:** If the eclipse is especially dark, photograph the background of stars and planets contrasted against the totally eclipsed moon. Equipment required for this type of record may be as simple as a tripod–mounted camera with a 50 mm lens, since field of view is the major ingredient.

Such photography may also allow you to capture a star about to be occulted or just emerging from occultation by the totally eclipsed moon. If you are at the right location, either by chance or design, you might have an opportunity to photograph a grazing occultation during totality!

For long exposures or relatively small fields of view, you will need to use a clock–driven mount.

•**Multiple exposures:** An impressive way to record a partial or total lunar eclipse is through multiple exposure. The steps are the same as those for solar eclipses. One difference is exposure length during totality; if totality is very dark, you will need an exposure so long that image trailing can be a problem. Use a spare clock–driven equatorial mount with the drive off for all but the exposure taken at totality.

Archiving your photographs. Preserving and storing the best of your eclipse photographs is an important consideration. Slides are best stored in metal or plastic slide trays, plastic pocket protector sheets which fit in three–ring binders (*not* polyvinyl chloride), or in projection trays in boxes. Plastic pocket protector sheets are also available in

Photograph 17-7 A demonstration of what can be done with fairly simple equipment: an unguided one second exposure taken through a Celestron Comet Catcher on Kodachrome of the 28 November 1993 lunar eclipse just prior to third contact. Unless the telescope and camera are clock–driven, exposures much longer in duration would result in the image being "smeared" due to the earth's rotation. *Photograph taken by Milt Hays.*

a variety of sizes for photographic prints, and traditional photo albums work well. Negatives should be stored in the plastic sleeves photo processors return them in, and should be carefully coded to prints for easy location when duplicates or enlargements are wanted. You might also consider having all your slides and/or negatives put on photo CD.

Documenting each slide, print and negative with date, time and exposure information is a critical step, and should be taken immediately, while the information is fresh in your mind. Avoid writing on the backs of prints; the words bleed through or the pen pressure can damage the

Photograph 17-8 Multiple exposures of the partial lunar eclipse phases can produce a nice eclipse series, even when clouds interfere as in this multiple exposure of the 6 July 1982 lunar eclipse. Note the bright stars. *Photograph taken by Mike Reynolds.*

finish. A small code number can be written on the back edge, if necessary, and indexed to a computer or card file; a consecutive numbering system, using the last two digits of the year as a prefix, works well for slides as well as prints. A similar code can be used to manage negatives. Computer programs are available that print out personalized stick-on labels for slides and the backs of prints.

In time, even the best-cared-for prints and negatives sustain damage and color dyes fade. Recent technologies offer excellent solutions to these problems. Simple video camera attachments allow prints and slides to be transferred to video tape (but consider the stability of video tape); if you don't have the necessary equipment, most commercial photo processors offer this service. Another service offered by many dealers converts your negatives and slides to photo compact disks (CD's). Once written to CD-ROM, your home computer and image-processing software allows you to enhance your images in exciting new ways!

Summary. Our recommendations for eclipse photography are as follows:

- PRACTICE in advance. This includes set-up, focus, exposure, tracking, loading film, and all the things that cannot happen but will happen! Use the moon as a test target.
- Carry spare batteries for your camera, data back, auto winder, and clock drive, beyond your normal assortment of tools and supplies (such as duct tape and extra hardware) for your camera system(s), lenses and telescope. Understand your photographic system's strengths and weaknesses.
- Carry a spare camera body, if possible. There are many horror stories about cameras jamming right at the beginning of totality! Many seasoned total solar eclipse observers will not only take several cameras to an eclipse but will take photographs with each just in case something unknown happens to one of them.
- Protect your equipment prior to and during the eclipse. This may include protection from weather elements like blowing sand and dust, moisture, and temperature (heat or cold). Remember that in extremely cold situations camera and clock drive batteries might fail and camera lubricants might freeze and film has the tendency to become brittle at extremely cold temperatures.
- Bracket exposures wherever possible.
- Shoot one ordinary scene on the roll of film at the beginning (especially if you are using slide film). Commercial film processors have cut images in half because they didn't recognize where one frame ended and the next one began.
- Be as careful archiving your photographs as you were taking them.
- Don't forget to look at the eclipse! Don't become so intent on photographing the event that you don't enjoy one of nature's most wonderful spectacles.

Many good articles in periodicals, such as **The Journal of the Association of Lunar and Planetary Observers, Astronomy** and **Sky & Telescope**, include photographs with exposure information. We highly recommend that you read about the eclipse experiences and recommendations of others as well.

References

Berry, Richard. *Bring On The Photo CD Revolution*. **Astronomy** (July, 1994). Waukesha, WI: Kalmbach Publishing Co., 1994.

Brasch, Klaus. *Sky Testing ScotchChrome 800/3200P Film*. **Sky & Telescope** (February, 1994). Cambridge, MA: Sky Publishing Corporation, 1994.

di Cicco, Dennis. *Where Are the Ambassadors? (Astrophotographers: a dying breed?)*. **Sky & Telescope** (April, 1994). Cambridge, MA: Sky Publishing Corporation, 1994.

Schur, Chris. *Getting Focused on Sharper Photos*. **Astronomy** (July, 1993). Waukesha, WI: Kalmbach Publishing Co., 1993.

18

Electronic Eclipses – Video and CCD Imaging

Introduction. Video and charge–coupled device (CCD) imaging have begun to make major inroads on traditional photographic emulsions among amateurs and professionals seeking to record astronomical events and objects for later study.

Electronic imaging offers observers opportunities for nearly instantaneous inspection of eclipse results, as well as a continuous recording of the event. CCD images also allow accurate photometry of eclipses. In addition, observers making video recordings can record an audio track during the event (a caution here: be careful of what you say during the eclipse if you want to later play the tape for others!). The audio track can also be used to record WWV (or other) time signals in order to precisely time eclipse events.

Video imaging. Since the 1980's, video cameras have been readily available on the market. These cameras are available in a variety of tape formats, from professional 3/4–inch video to super 8 mm. They utilize a type of light–sensitive electronic chip similar to that of the charge–coupled devices. While off–the–shelf video cameras are not good for recording low light level events and objects (such as deep sky objects or planets at high focal ratios), they are excellent for lunar and solar eclipses and, with a medium–size telescope, can also record the eclipses and shadow transits of Jupiter's satellites.

Equipment. Most off–the–shelf video cameras come with features that allow auto or manual focus and zoom from wide field to telephoto, as well as other more advanced features. If you do not yet own a video camera, there is a plethora of models on the market sporting features especially useful for successful recording of eclipses.

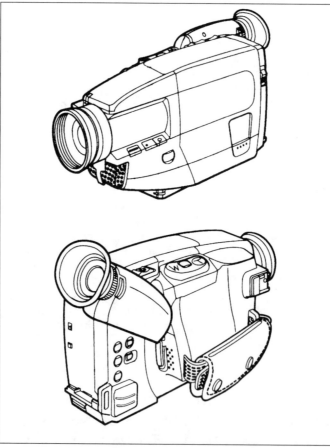

Figure 18-1 The Minolta MASTER 8–778 8 mm camcorder. This camcorder, like other commercially-available camcorders, features 8 power zoom, 1 lux low–light level recording and electronic image stabilization. Other features include fade–in/fade–out, character generator and manual or auto focus. *Illustration © Minolta Corporation; reproduction is made with the permission of Minolta Corporation.*

• **Tape format:** Most off–the–shelf video cameras will either be in a standard VHS, VHS–C (compact) or 8 mm format. There are sound reasons for considering each format, depending upon your preferred replay options and the amount of money you want to invest.

Of the three formats, only 8 mm uses a metallic tape, which produces a better quality picture. In addition 8 mm has the potential to hold up better over time. However you cannot play 8 mm in most VCR's and therefore must use the camcorder itself for replay; some 8 mm and dual–deck VCR's are now on the market.

• **Exposure setting:** Many video cameras include the option of automatic or manual exposure. Some operate through varying a lens diaphragm or iris, whereas others utilize a variation in shutter speed. Some camcorders can shoot as fast as 1/10,000th of a second, valuable for extremely short exposure requirements.

It is important that an eclipse chaser's camera

have a variable exposure feature for both total lunar and total solar eclipses. Sometimes an automatic exposure camera can be "tricked" by adjusting the shutter speed.

• **Lux ratings:** Lux is a unit of illumination. A camcorder's lux rating refers to it's light sensitivity. Many video cameras advertise lux ratings as low as 1 lux, helpful when trying to record faint details. However if the camera is on automatic exposure, these details might be "washed out" by nearby brighter features.

• **Battery/portability:** One of the most important features is the freedom from dependency on an alternating current (A.C.) power source. Be sure to look for a battery pack that will provide enough power to last through the eclipse. Extra battery packs, a battery charger and an international voltage converter are other essentials.

• **Lens systems:** Most offer a zoom feature from wide angle to telephoto, usually 8:1, 10:1, or 12:1. Some camcorders have a digital zoom feature of up to 24:1, but the results are grainy. Check this feature to determine the maximum zoom focal length at the telephoto setting; most are in the range of 60 to 80 mm (producing a fairly wide field of view) which is not sufficient to record a lunar or solar image of acceptable size, except for the corona during a total solar eclipse. Flexibility is usually restricted by the fact that off–the–shelf video camera lenses are not easily removable, as are the lenses of commonly–owned 35 mm single lens reflex cameras. This is due to the electronics built into the video camera's lens to control auto focus and zoom features.

• **Telextenders:** There is an accessory that can eliminate the image size problem. Telextenders, which screw onto the front of the camcorder's normal lens, can significantly increase the effective image size. These telextenders are available in different magnifications, from 1.5 times up to 5 times or more the original image size. Beware: the expense of such telextenders can also be high, and they can degrade the image quality! However, video is more forgiving than film. One may pay $100 or more for a video telephoto and be disappointed in the resulting image. Vignetting can also result from using a lens telextender if the camera lens is set to a focal length that is too short. Video telextenders are available at camera and electronics stores as well as from some astronomical equipment distributors. Shop as carefully for this accessory as for the initial video camera purchase itself.

• **Afocal systems:** An alternative to the telextender is to employ an afocal system. Afocal systems match the camera's lens to the complete optical system (objective and eyepiece). A monocular or a finder scope makes an ideal system to match to the video camera. Lower power is recommended; too high a magnification will produce an unsatisfactory image (a large, washed–out image with vibrations greatly exaggerated, and an overall poor quality). We recommend that you mount the entire system (optical system–video camera) to a base plate that will allow you to optically center the system as well as assure that the system will not move during the eclipse. You will also want to block all light that can enter between the two components; such a light block is easy to make from material as common as cardboard or black felt.

• **Other features:** These may include time–lapse exposure settings (e.g., one exposure every 30 seconds), date– and time–labeling, fade–in/fade–out, and subtitling. One feature that is *not* useful is automatic focusing. Set the camera to "manual focus" and focus the lens at infinity.

You may also want to carry along a separate monitor to view your images directly. In addition to seeing your images much more clearly, this will allow others to enjoy the eclipse through the video camera's lens with you.

Mounting the video camera. Since camcorders come equipped with a standard photo thread mount in the base of the camera, you have several options for mounting the camera instead of hand–holding it. These include piggy–backing on a telescope, mounting directly on an equatorial mount, and attaching the camera to a tripod. Unfortunately the tripod's screw thread mount is usually fastened to the flexible plastic case of the camera; the camera should be fastened down at more than one point using clamps, straps, etc., to minimize vibrations.

• **Hand–holding:** A reasonable video may be obtained by hand. Problems include the small field of view if a telephoto lens accessory is being used, and the difficulty of holding the camera steady for a long period of time. This is especially true if the eclipse is near the zenith, or if you tend to do a lot of jumping with joy during totality. This approach should probably be reserved for use as a last resort!

• **Camera tripod:** A high quality, heavy–duty photographic or video camera tripod will provide a steady mount for the video camera. The one drawback: the eclipse chaser or family member must continuously adjust the tripod to keep the target in the camera's field of view. An adjustable slow–motion tripod–mounted pan head would be a very useful accessory.

• **Piggy–backing units:** If the video camera is light–weight (as are the VHS–C and 8 mm systems) or if the telescope and mount are heavy duty, one can actually piggy–back the video camera as part of the system. Be careful not to bump the telescope; remember that the video will pick up these wiggles and one doesn't want to return from a clear total solar eclipse expedition with poor video due to vibrations! Some camera systems now include a feature called *image stabilization* that eliminates some wiggles by compensating for the vibrations.

• **Equatorial mounts:** A clock–driven equatorial mount will allow you to set the video camera on automatic, leaving you free to do other things during the eclipse. Most video cameras are fairly small units (in comparison to telescopes) so a light–weight equatorial mount can suffice if packing a lot of equipment is a concern. A good equatorial mount will be useful if you decide to employ an afocal system. A heliostat–type arrangement could also be used with the camcorders.

Charge–coupled devices. Charge–coupled devices have recently made major inroads into both professional and amateur circles. Like the video camera, the CCD is a sold–state light detector, in the form of a chip, which detects photons (light) at thousands of photosites. At these photosites, the light frees electrons that are, in turn, moved to an amplifier by charge–coupling. The output of the amplifier is digitized from an analog signal.

CCD's are readily available to the amateur astronomer in a great variety of configurations. These configurations dif-

shadow of a Galilean satellite across Jupiter's cloud tops or a satellite entering or leaving Jupiter's shadow, the CCD is probably unparalleled.

The quality of the CCD image is unsurpassed; the ability to process an image with one's computer allows the amateur to produce professional–quality imaging that exceeds the capability of photographic films.

Eclipse observations using electronic imaging. Procedures vary somewhat depending upon whether an eclipse is lunar or solar.

•**Lunar eclipses:** Video and CCD imaging of partial and total lunar eclipses can produce some spectacular results. The CCD allows the observers to conduct not only imaging but also photometry of the eclipse due to the light sensitivity and tonal range of the CCD. The CCD is very effective at imaging varying brightness levels.

Video cameras will have difficulty automatically adjusting to the vast differences in the amount of light coming from the eclipsed moon even during totality. We recommend that a variety of "exposure lengths" or shutter speeds (actually, the amount of light allowed to enter the camera) be tried instead of attempting to vary the exposure through the iris of the video camera.

CCD's and video cameras should also be used to image medium and long–duration Lunar Transient Phenomena.

•**Solar eclipses:** As with any type of solar observation, sufficient filtering is necessary during partial phases. A denser–than–usual solar filter

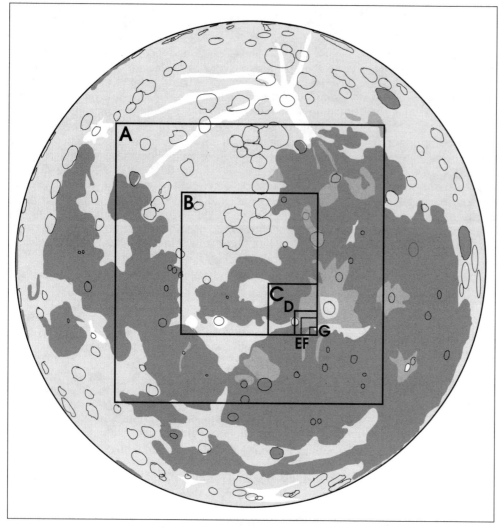

CCD Lunar Image Coverage
TC211 Sensor, 192 x 165 pixels, 2.64–mm square array size

Sample Telescope	Size Covered	Area Covered	Pixel Size
A Comet Catcher (f/3.64 @ 51 cm FL)	1,994 X 1,994 km	3,975,000 sq km	11.20 km
B Celestron C–90 (f/10 @ 100 cm FL)	1,014 X 1,014 km	1,028,000 sq km	5.70 km
C Celestron C–11 (f/10 @ 279 cm FL)	363 X 363 km	131,600 sq km	2.04 km
D Celestron C–11 (f/21.4 @ 598 cm FL)	170 X 170 km	28,750 sq km	0.95 km
E Celestron C–11 (f/29 @ 810 cm FL)	125 X 125 km	15,650 sq km	0.70 km
F 51 cm Refractor (f/16.8 @ 853 cm FL)	119 X 119 km	14,110 sq km	0.67 km
G 51 cm Refractor (f/36 @ 1826 cm FL)	55.5 X 55.5 km	3,081 sq km	0.25 km

Figure 18-2 An illustration of CCD lunar image coverage with a TC211 sensor, 192 X 165 pixels, 2.64 mm square array size. *Illustration by John Westfall.*

fer in terms of the chip size (often 165 x 192 photosites) and their dynamic range (commonly 8-bit, 12-bit, or 16-bit) as well as in their noise reduction capabilities and overall chip quality.

The difficulty with using CCD imaging for lunar and solar eclipse work is the tiny field of view. Since the chip size is so small the field of view is accordingly small. One could couple a CCD with a short focal length system (say a 200 mm focal length achromat). However, for imaging the may be necessary due to the light–sensitive nature of the CCD chip (although it is hard to physically damage a CCD chip).

A variety of exposure lengths during totality (minus the filter!) will bring out the various features, from the inner corona and prominences to the graceful outer corona and solar streamers. The appearance and reappearance of Baily's beads and the diamond ring effect will effectively stop down the system if you are recording on automatic,

Photograph 18-3 VHS image of the partial phase of the total lunar eclipse, 17 August 1989. Video was taken through a Celestron Comet Catcher with a Sony Black & white camera. South is in the lower left corner. *Video taken by John Westfall.*

thus losing the beauty of the beginning (or ending) of totality.

CCD imaging of the changing coronal features is effective due to the fact that the CCD differentiates the different light levels well.

The video camera can also be used to record other spectacular eclipse phenomena, from horizon colors to shadow bands to even the movement of the shadow itself across the sky. A fish-eye lens or wide field attachment or pointing the video camera at a hemispherical hubcap can give all-sky coverage. Observers have even reported using the video camera to record animal (including human) responses to totality.

•**Jupiter satellite eclipses:** Using a video camera with a medium-size 15 to 20 cm (6 to 8 in) telescope can record Jupiter's Galilean satellites entering or leaving the planet's shadow, as well as the satellites' shadows crossing Jupiter (creating a solar eclipse for those Jovians who choose to travel to the center line!). If you can remove your video camera's lens, do so and place the camera chip at the focal plane of the telescope. If not, set the camera lens to infinity and point it through the telescope eyepiece. A true CCD will provide better results for those observers attempting to record Jupiter satellite eclipses.

Summary. The following are our recommendations for eclipse video imaging:

- Carry back-up (and charged) batteries and video tapes.
- Use a front-of-the-lens telephoto but don't use a magnification so high that it causes color changes, easily seen vibrations, degraded quality of images, or excessive vignetting.
- Make certain that during a total solar eclipse you can record the outer corona and any streamers.
- Take the video camera out of the automatic focus mode; set the camera on infinity focus.
- Be familiar with the exposure mode and vary it during the event (like bracketing during conventional astrophotography). Otherwise you will record only one type of image.
- Finally, PRACTICE in advance. This is easier to do with a video camera than with a photographic system—you get immediate results. The full moon makes a great target for both lunar and solar eclipse practice; a moon with visible earthshine makes for great total lunar eclipse practice.

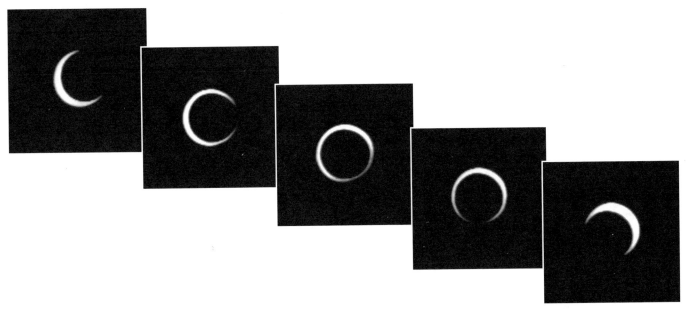

Photograph 18-4a–e A composite image series of the 10 May 1994 annular eclipse VHS images. Video was taken with a Minolta Master Series-V 18R VHS camcorder and a 3X telextender. *Video taken by Jeremy Reynolds.*

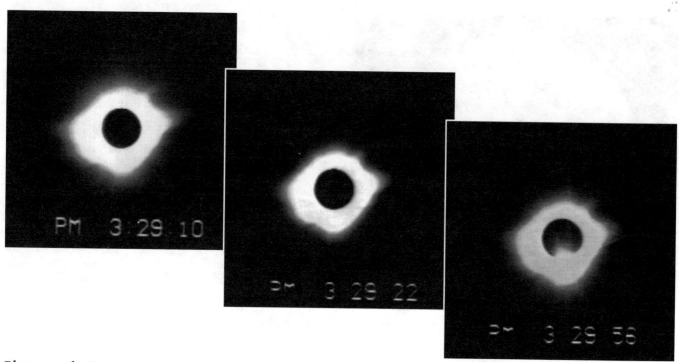

Photograph 18-5a–c A series of the 11 July 1991 total solar eclipse VHS images as seen south of Puebla, Mexico. Video was taken with a Minolta camcorder and a 3X telextender. Note the time–labeling. *Video taken by Mike Kazmierczak.*

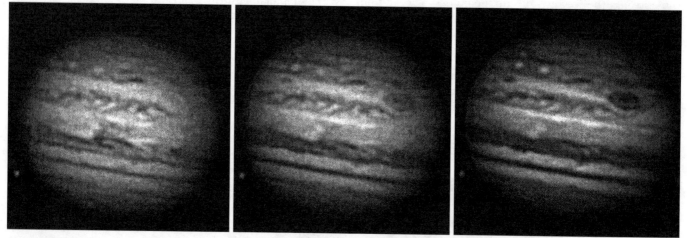

Photograph 18-6a–c CCD images of Jupiter and Ganymede off NE limb (left of Jupiter), 15 May 1993. The left image, 1.5 second exposure, was taken in red light. The middle image, 2.3 second exposure, was taken in green light. And the right image, 4.3 second exposure, was taken in blue light. A LYNXX COD CCD camera on a 41 cm f/6 Newtonian reflector at an effective focal length of f/27 was used to take these images. Imagepro software was used to process the images. Three images in specific colors can be combined to produce a color image. *CCD images taken by Don Parker.*

References

Berry, Richard. **Choosing and Using a CCD Camera**. Richmond, VA: Willmann–Bell, Inc., 1992.

Berry, Richard. **Introduction to Astronomical Image Processing**. Richmond, VA: Willmann–Bell, Inc., 1991.

Buil, Christian. **CCD Astronomy: Construction and Use of an Astronomical CCD Camera**. Willmann–Bell, Inc., 1991.

Appendices

The totally–eclipsed sun as it would be seen by one of the many cruise ships which carry eclipse enthusiasts into the path of the moon's shadow at sea. *From a painting by astronomical artist Vina Loper.*

Appendix 1

Recommended Books

Association of Lunar and Planetary Observers. **Lunar Observer's Manual**. Member–provided service to train amateurs to make scientifically useful lunar observations; contact the A.L.P.O.

Bishop, Roy L., ed. **Observer's Handbook**. Toronto: The Royal Astronomical Society of Canada. Annual volume of astronomical data and tables, including profiles of the year's eclipses.

Brewer, Bryan. **Eclipse, 2nd ed**. Seattle, WA: Earth View, Inc., 1991. Concise overview of solar and lunar eclipses for beginning to intermediate amateurs; many illustrations and photographs.

Espenak, Fred. **Fifty Year Canon of Solar Eclipses: 1986–2035**. Cambridge, MA: Sky Publishing Corporation, 1987. Compilation of solar eclipse data and maps.

Espenak, Fred. **Fifty Year Canon of Lunar Eclipses: 1986–2035**. Cambridge, MA: Sky Publishing Corporation, 1989. Compilation of lunar eclipse data and maps.

Graham, Francis and Westfall, John. **Lunar Eclipse Handbook**. Lunar Press, 1990. Association of Lunar and Planetary Observers member–provided guide for amateurs conducting useful lunar eclipse observations; contact the A.L.P.O. for information.

Harris, Joel and Talcott, Richard. **Chasing the Shadow**. Waukesha, WI: Kalmback Publishing, 1994. An historical and observational overview of solar eclipses.

King–Hele, Desmond. **Observing Earth Satellites**. New York: Van Nostrand Reinhold Company, 1983. Guide to observing and photographing artificial satellites.

Littmann, Mark and Willcox, Ken. **Totality**. Honolulu, HI: University of Hawaii Press, 1991. Historical, mechanical and observational aspects of solar eclipses with a special overview of the 11 July 1991 total solar eclipse.

Liu, Bao–Lin and Fiala, Alan D. **Canon of Lunar Eclipses 1500 B.C.–A.D. 3000**. Richmond, VA: Willmann–Bell, Inc., 1992. Detailed information on 4,500 years of lunar eclipses.

Maag, Russell C., Sherlin, Jerry M., and Van Zandt, Rollin P., ed. **Observe and Understand the Sun**. Washington, D.C.: Astronomical League, 1976. Introductory observing guide to the sun; many helpful hints.

Meeus, Jean. **Transits**. Richmond, VA: Willmann–Bell, Inc., 1989. Details on planetary transits including those visible from planets other than the earth.

Meeus, Jean., Grosjean, C. C., and Van Der Leen, W. **Canon of Solar Eclipses**. Oxford: Pergamon Press, 1966. A compendium of solar eclipse data and path maps, 1898–2510.

Ottewell, Guy. **The Astronomical Calendar**. Greenville, SC: Astronomical Workshop, Department of Physics, Furman University. Annual almanac of celestial events including all eclipses; beautifully illustrated.

Ottewell, Guy. **The Astronomical Companion**. Greenville, SC: Astronomical Workshop, Department of Physics, Furman University, 1981. Lavishly illustrated companion to the annual **Astronomical Calendar** which explains many technical aspects of eclipses and orbital motions in simple terms.

Ottewell, Guy. **The Understanding of Eclipses**. Greenville, SC: Astronomical Workshop, Department of Physics, Furman University. An overview of the mechanics of eclipses including a detailed look at eclipses of 1989 through 1991; beautifully illustrated.

Pasachoff, Jay M. and Covington, Michael A. **The Cambridge Eclipse Photography Guide**. Cambridge, Great Britain: Cambridge University Press, 1993. An overview of eclipse mechanics, observing eclipses and eclipse photography; a beautiful color eclipse plate section.

von Oppolzer, Theodor R. **Canon of Eclipses**. New York: Dover Publications, 1962. Contains data and maps for eclipses, 1207 B.C.–2161 A.D.

U. S. Government Printing Office. **The Astronomical Almanac**. Washington, D.C.: U. S. Government Printing Office. Annual volume of astronomical data and tables, including profiles on the year's eclipses.

Zirker, J. B. **Total Eclipses of the Sun**. New York: Van Nostrand Reinhold Company, 1984. Concise examination of solar eclipses for intermediate amateurs.

Appendix 2

Recommended Periodicals and Software

Periodicals:

The following listing has been compiled to offer the eclipse observer guidelines for carrying out observational projects; it is not intended to be comprehensive. References, with the exception of those from the **J.A.L.P.O.**, have been annotated and chosen from the two–and–a–half years immediately preceding publication of this edition of **Observe Eclipses** to simplify access. **J.A.L.P.O.** article titles are largely self–descriptive.

From **The Journal of the Association of Lunar and Planetary Observers (J.A.L.P.O.): The Strolling Astronomer**,
P.O. Box 143, Heber Springs, AR 72543.

Lunar eclipses:
Staff. A Total Eclipse of the Moon For The Americas: 1993 Nov. 29. Vol. 37, No. 2 (October, 1993).
Staff. Observing the Next Total Lunar Eclipse: 1989 Feb. 20. Vol. 33, No's. 1–3 (January, 1989).
Westfall, John E. Photoelectric Photometry of the 1983 June 25 Partial Lunar Eclipse. Vol. 32, No's. 9–10 (August, 1988).
Westfall, John E. The Total Lunar Eclipse of July 6, 1982: A Dark and Asymmetric Umbra. Vol. 31, No's. 9–10 (July, 1986).
Westfall, John E. Three–Color Photometry: Penumbral Lunar Eclipse, May 15, 1984 U.T. Vol. 30, No's. 9–10 (August, 1984).

Lunar transient phenomena:
Cameron, Winifred Sawtell. Results From The LTP Observing Program For The Association of Lunar and Planetary Observers (ALPO). Part 1: Vol. 27, No's. 9–10 (March, 1979); Part 2: Vol. 27, No's. 11–12 (June, 1979).
Darling, David O. and Weier, David D. Evidence of an Apparent Dust Levitation or Outgassing in the Crater Tycho. Vol. 37, No. 3 (February, 1994).
Westfall, John E. A Comprehensive Catalog of Lunar Transient Phenomena. Vol. 27, No's. 9–10 (March, 1979).

Solar eclipses:
Glenn, William H. Sky Color and Darkness at the Total Solar Eclipse, 10 July. 1972. Parts I, II and III. Vol. 28, No's. 1–2 (October, 1979), No's. 3–4 (January, 1980), and No's. 5–6 (April, 1980).
Graham, Francis G. A Summary of Observations of the 1989 Mar 07 Partial Solar Eclipse. Vol. 34, No. 3 (July, 1990).
Hill, Richard E. Observations of the May 30, 1984 Annular Solar Eclipse. Vol. 30, No's. 11–12 (November, 1984).
Staff. The Great Total Solar Eclipse of July 11, 1991. Vol. 35, No. 2 (June, 1991).
Staff. The Annular Solar Eclipse of 1994 May 10. Vol. 37, No. 3 (February, 1994).

Eclipses of planetary satellites:
Bulder, Henk J. J. Using a CCD Video Camera for Galilean Satellite Eclipse Timings. Vol. 36, No. 4 (February, 1993).
Westfall, John E. Galilean Satellite Eclipse Timings: The 1989/90 Apparition. Vol. 36, No. 2 (July, 1992).
Westfall, John E. Galilean Satellite Eclipse Timings: 1986/87 Report. Vol. 32, No's. 11–12 (October, 1988).

From **Sky & Telescope** magazine,
Sky Publishing Corporation,
49 Bay State Road, Cambridge, MA 02138.

Lunar eclipses:
di Cicco, Dennis. Forecasting a Lunar Eclipse. June, 1994. Discussion of efforts to predict the darkness of a lunar eclipse from data on atmospheric conditions.
Jawad, Ala'a H. How Long Is A Lunar Month? November, 1993. Interesting discussion of the variable length of the synodic month.
MacRobert, Alan M. November's Eclipse of the Moon. November, 1993. Prospects and projects for the 28 November 1993 total lunar eclipse.
MacRobert, Alan M. The May 24th Partial Lunar Eclipse. May, 1994. Preview and observing hints.
Mallama, Anthony and Caprette, Douglas. Anatomy of a Lunar Eclipse. September, 1993. Discusses a new technique for processing a CCD image of a lunar eclipse to leave only the effect of earth's shadow.
O'Meara, Stephen James. A "Diamond–Ring" Lunar Eclipse. March, 1994. Observations of the unevenly illuminated total lunar eclipse of 28–29 November 1993.
O'Meara, Stephen James. June's Colorful Lunar Eclipse. November, 1993. Observations of the 4 June 1993 total lunar eclipse.
O'Meara, Stephen James. The Night the Moon Disappeared. April, 1993. Observations of the extremely dark total lunar eclipse of 9–10 December 1992.
O'Meara, Stephen James. Making Sense of November's Perplexing Lunar Eclipse. June, 1994. Analysis of apparently conflicting observations of the 28–29 November 1993 eclipse; interesting photographs and artwork from observers.

Solar eclipses:
Anderson, Jay. Seeking Clear Eclipse Skies. November, 1993. Prospects for the 3 November 1994 total solar eclipse.
Harris, Joel K. November 3, 1994: Waiting For Totality. November, 1993. Weather prospects and travel tips for the total solar eclipse of 3 November 1994.
MacRobert, Alan M. A Partial Solar Eclipse on May 21st. May, 1993. Features viewing safety considerations.
MacRobert, Alan M. A Solar Eclipse for Everyone. May, 1994. Preview, observing and photography hints.

Schaefer, Bradley E. *Solar Eclipses That Changed the World.* May, 1994. Impacts of solar eclipses on historical events.

From **Astronomy** magazine,
Kalmbach Publishing Co.,
21027 Crossroads Circle, P. O. Box 1612, Waukesha, WI 53187.

Lunar eclipses:

Bruning, David. *November's Colorful Eclipse.* April, 1994. Summary of observations and photography of the 28/29 November 1993 total lunar eclipse.

Dyer, Alan and Talcott, Richard. *Eclipse Over America.* November, 1993. Examines observing and photography prospects for the 28/29 November 1993 total lunar eclipse.

Dyer, Alan and Talcott, Richard. *When the Moon Disappears.* December, 1992. Prospects and photo tips (including video), for the 9/10 December 1992 total lunar eclipse.

Staff. *Viewing the Lunar Eclipse.* May, 1994. Reports, photo and binocular sketch of the 28/29 November 1993 total lunar eclipse.

Solar eclipses:

Anderson, Jay and Willcox, Ken. *Return to Darkness.* November, 1993. Travel planner for the total solar eclipse of 3 November 1994.

Bruning, David. *One Day On The Sun.* January, 1992. An explanation for the unexpectedly asymmetrical appearance of the corona of the 11 July 1991 total solar eclipse.

Talcott, Richard. *Ring of Fire.* April, 1992. Prospects for the 4 January 1993 annular solar eclipse.

Talcott, Richard. *Ring of Fire.* May, 1994. How to observe and photograph the annular eclipse of 10 May 1994.

Software:

The following listing includes software advertised from January, 1993 to June, 1994 as being capable of calculating or illustrating eclipses of the sun and moon. Quoted descriptions are from company advertisements; potential purchasers should not consider these entries as recommendations. The program listing is followed by the address and telephone numbers of the companies.

Computer Canon of Lunar Eclipses 1500 BC to AD 3000. Willmann–Bell, Inc. Optional software to accompany the book **Canon of Lunar Eclipses 1500 BC to AD 3000** by Liu and Fiala. "Circumstances of 10,990 lunar eclipses can be interactively displayed on... screen, saved... or printed.... Graphic displays."

Distant Suns. Virtual Reality Laboratories, Inc. "Reproduces eclipses...." IBM & compatibles.

EclipseMaster Plus™. Zephyr Services. "Computes data for solar & lunar eclipses (no maps). All aspects included.... Input saros number and get for each eclipse in the series." IBM & compatibles.

MoonTracker EGA™. Zephyr Services. "High resolution lunar eclipses; displays areas of visibility on a world map." IBM & compatibles.

Night Sky, The. Andromeda Software, Inc. "The eclipse portion... will predict and plot Solar and Lunar eclipses on a map of the earth. The Lunar eclipse drawing will show the path of the moon across the Earth's shadow...." IBM & compatibles.

Observer's Companion, The. Arc Science Simulations. "Calculating almanac finds past and future eclipses...." Requires **Dance of the Planets** to run. **Dance of the Planets.** Comprehensive planetarium program. IBM & compatibles.

PC–SKY. CapellaSoft. "Preview eclipses; view historical celestial events." IBM & compatibles.

PEEP: Planetary Event and Eclipse Predictor. CEB Metasystems, Inc. "...calculates the circumstances of solar and lunar eclipses... from 4,713 B.C. to 10,000 A.D." IBM & compatibles.

Solar. Andromeda Software, Inc. "...solar eclipse prediction program... for determining the geographical limits... for every solar eclipse from March 7, 1951 through May 9, 2032...." IBM & compatibles.

SunTracker Premier™, SunTracker Pro™, SunTracker Super™, and **SunTracker Plus™.** Zephyr Services. "Plots solar eclipses on variable scale world maps, plus numerical data." IBM & compatibles.

TotalEclipse™. Zephyr Services. "An integrated package for investigating solar and lunar eclipses from 4712 B.C. to 9999 A.D. that generates... dates, times... magnitudes and... graphical displays." IBM & compatibles.

Voyager II™. Carina Software. A program that produces star, constellation and object charts for the observer as required. Macintosh.

Software distributors: Contact the following for current catalogs and prices for the software listed above. Information was current in 1994, but is subject to change. Software is also available through a number of additional distributors and vendors.

Andromeda Software, Inc., P.O. Box 605, Amherst, NY 14226–0605. (716) 691–6731 fax orders.

Arc Science Simulations, P.O. Box 1955A, Loveland, CO 80539. (303) 667–1168 information, (303) 667–1105 fax orders, (800) 759–1642 orders only. CapellaSoft, P.O. Box 3964, La Mesa, CA 91944. (619) 460–8265 information and orders, (619) 463–6067 fax orders.

Carina Software, 12919 Alcosta Boulevard, Suite 7, San Ramon, CA 94583. (510) 352–7328.

CEB Metasystems, Inc. 1200 Lawrence Drive, Suite 175, Newbury Park, CA 91320.

Willmann–Bell, Inc. P.O. Box 35025, Richmond, VA 23235. (804) 320–7016 information and orders, (804) 272–5920 fax orders.

Virtual Reality Laboratories, Inc., 2341 Gandor Court, San Luis Opispo, CA 93401. (805) 545–8515 information, (805) 781–2259 fax orders, (800) 829–8754 orders. (805) 499–0958 information, (805) 375–6097 fax orders, (800) 232–7830 orders.

Zephyr Services, 1900 Murray Ave., Dept. B., Pittsburgh, PA 15217. (412) 422–6600 information, (412) 422–9930 fax orders, (800) 533–6666 orders only.

Glossary of Selected terms

Afocal system – Method of telescopic photography where the camera's lens is placed against the telescope's eyepiece to produce an image.

Annular eclipse – A type of solar eclipse in which the moon's angular diameter is too small to completely cover the sun, leaving a ring of light called an "annulus" surrounding the lunar disk.

Annular–total eclipse – A type of solar eclipse in which the moon's angular diameter is so close to the angular diameter of the sun that along most of the eclipse path an annular eclipse is seen while a total eclipse is visible along part of that path.

Ascending node – That point along the orbital path of the moon (or a planet) where it crosses the ecliptic moving north.

Baily's beads – A phenomenon of total and some annular solar eclipses in which sunlight continues to shine through valleys and other depressions on the lunar limb after the moon has otherwise completely covered the sun's face.

Black drop effect – During transits, the planet's limb appears to delay its separation from or hasten its reunion with the sun's limb. May also occur during an annular solar eclipse second and third contact.

CCD – Charge–Coupled Device; a solid state light detector which detects photons at thousands of photosites. CCD's are manufactured in the form of a chip.

Central eclipse – A solar eclipse which is either total, annular, or annular–total, in which the moon passes centrally across the sun's face. May also refer to a total lunar eclipse, in which the moon passes centrally through the earth's shadow.

Chromosphere – The layer of the sun's atmosphere closest to the surface of the sun.

Corona – The sun's atmosphere beyond the chromosphere, which becomes visible at the time of a total solar eclipse.

Descending node – That point on the orbital path of the moon (or a planet) where it crosses the ecliptic }moving south.

Diamond ring – An effect of total solar eclipses caused by sunlight shining from one small region of the sun's limb which the moon has not yet covered or has just uncovered.

Eclipse – An apparent dimming or extinction of light coming from one heavenly body by another body.

Eclipse seasons – Periods when eclipses are possible, occurring at intervals roughly six months apart, when the line of nodes of the moon's orbit is aligned with the sun and the moon is near one of its nodes. Three such seasons are possible in a given calendar year.

Eclipse year – The time it takes for the same node of the moon's orbit to make consecutive alignments with the sun; 346.62 days, or 341.64 degrees of solar ecliptic travel due to the earth's orbital motion.

Ecliptic – The sun's apparent path through the heavens.

Flash spectrum – The sudden change in the sun's spectrum at second contact when the dark absorption lines of the photosphere are replaced by the bright emission lines of the chromosphere.

First contact – The moment that the moon first appears on the edge of the sun in a solar eclipse, or when the earth's umbral shadow first appears on the moon's limb in a lunar eclipse; signifies the beginning of an eclipse.

Fourth contact – The moment when the moon leaves the solar disk in a solar eclipse, or when the lunar disk leaves the earth's umbral shadow in a lunar eclipse; signifies the end of an eclipse.

Grazing occultation – The passage of one smaller celestial object tangent to a larger object. In the case of the moon and a star, the moon's irregular limb can create multiple disappearances and reappearances of the star. Information regarding the shape of the moon's limb, the position and possible multiplicity of the star can be derived from the data.

Inex – A repetitive series of 780 eclipses lasting 23,000 years; each eclipse in the series recurs, after 358 synodic months, in the same longitude but in the opposite hemisphere's latitude.

Inferior conjunction – Refers to the alignment of a celestial body between the earth and the sun, as opposed to superior conjunction, when the body is on the opposite side of the sun from the earth.

ISO – refers to the speed or light sensitivity of photographic film. The higher the ISO number, the more sensitive the film. ISO (International Standard Organization) is often referred to by an older identification, ASA (American Standards Association).

Line of nodes – Imaginary line connecting the ascending and descending nodes of the moon's (or a planet's) orbit and intersecting the earth.

Lunar eclipse – The moon, on the opposite side of the earth from the sun, passes through the earth's shadow cone.

Lunar transient phenomena (LTP) – Short-lived flashes of light, glows, and patches of haze reported periodically as being observed on the moon by earth-based observers.

Magnitude – For a solar eclipse, the fraction of the sun's diameter obscured by the moon at greatest phase, measured along the common diameter; for a lunar eclipse, the fraction of the moon's diameter obscured by the earth's penumbra (for penumbral eclipses) or umbra (for umbral eclipses) at greatest phase, measured along the common diameter.

Metonic cycle – A 19-year interval after which the moon repeats the same series of phase changes on the same calendar date.

Negative Projection – A type of photographic system often used to increase the effective focal length of a telescope by placing a Barlow lens between the telescope's objective and the camera. A negative projection system is generally not as effective as a positive projection system.

Occultation – The passing of a large body in front of another, smaller, body thereby blocking off light from the latter for a short period of time.

Partial lunar eclipse – Occurs when the moon passes through the earth's shadow, but not centrally, leaving part of the moon outside the dark umbral shadow.

Partial solar eclipse – Occurs when the moon only covers part of the sun, due to its path being north or south of the sun's center as seen from the observing site.

Penumbral lunar eclipse – The moon passes through part of the earth's light secondary shadow, the penumbra, but misses the darker umbral shadow.

Penumbra – The earth's secondary shadow, within which partial sunlight falls.

Photosphere – The visible face of the sun; the solar surface.

Positive Projection – A type of photographic system often used to increase the effective focal length of a telescope by placing an eyepiece between the telescope's objective and the camera.

Prime focus – Method of telescopic photography where the telescope's lens or mirror serves as the camera's lens without use of an eyepiece or traditional camera lens.

Prominences – Jets of gas which erupt from the photosphere and which can be observed at the moment of total solar eclipse.

Regression of nodes – Slow clockwise rotation of the line of nodes of the moon's orbit by 19 degrees per year, with one complete rotation taking 18.6 years; makes possible a third eclipse season in a calendar year. Planet orbits experience similar regression.

Saros cycle – The period of time between repetitions of solar eclipses having the same general characteristics; 223 synodic months or 6585.32 days.

Second contact – Marks the beginning of annularity or totality for a solar eclipse, and the beginning of totality for a lunar eclipse. For an annular solar eclipse, it is the moment when the moon's trailing limb reaches the sun's disk; for a total solar eclipse, it is when the moon's leading edge completely covers the sun's disk. For a lunar eclipse, it is the moment when the moon is completely inside the earth's umbral shadow.

Shadow bands – Bands of shadow which are seen rapidly crossing the ground in the seconds just before and just after totality in solar eclipses.

Sidereal month – The time it takes the moon to make one complete orbit of the earth; 27.32 days.

Single Lens Reflex Camera or SLR – A camera system in which the image seen by the photographer is the same image that will strike the film plane. Most SLR's feature an interchangeable lens system, an important plus to the eclipse chaser.

Solar eclipse – The moon, on the same side of the earth as the sun, passes in front of the sun as seen from the earth.

Sunrise-sunset effect – brilliant red and orange horizon colors that often occur just prior to, during and just after a total solar eclipse for an observer in the path of totality.

Synodic month – The time it takes for the moon to repeat a previous phase, as from one new moon to the next; 29.53 days.

Telextender – An additional lens system that is placed between the camera body and the camera lens. The telextender increases the focal length of the camera lens.

Third contact – Marks the ending of annularity or totality for a solar eclipse, and the ending of totality for a lunar eclipse. For an annular solar eclipse, it is the moment when the moon's leading limb leaves the sun's disk; for a total solar eclipse, it is when the moon's trailing limb moves onto the sun's disk exposing the photosphere to view. For a lunar eclipse, it is the moment when the leading edge of the moon leaves the earth's umbral shadow.

Total lunar eclipse – The moon passes centrally or nearly so through the penumbral and umbral shadows of the earth.

Total solar eclipse – The moon passes directly in front of the sun and has an apparent angular diameter greater than the sun, thereby blocking off all of the sun's direct light.

Transit – The passing of one object (or its shadow) across the face of another, larger object, as with the planets Mercury or Venus transiting the sun or the shadow of one of Jupiter's moons transiting the face of Jupiter.

Umbra – The dark, central shadow cast by the earth.

Zodiac – A collective name for the twelve constellations which lay along the ecliptic and through which the sun and moon moves.

Appendix 4

Astronomical Organizations

The following astronomical organizations carry out eclipse–related observational research programs in which amateurs are encouraged to participate. They are always interested in receiving eclipse observations and photographs in their respective areas of interest. Courtesy mandates that all reports and inquiries to these organizations be accompanied by a self–addressed, stamped return envelope.

The Association of Lunar and Planetary Observers (A.L.P.O.)

Harry D. Jamieson, Membership Secretary
P.O. Box 143
Heber Springs, AR 72543

Founded in 1947, dues include a subscription to the quarterly *Journal of the Association of Lunar and Planetary Observers: The Strolling Astronomer*, which publishes observations, sketches and photographs by members. Separate Section Recorders offer observing manuals at nominal cost and training in their solar system disciplines, including solar and lunar eclipses. Meets annually, usually in the summer.

The Astronomical League (A.L.)

Ken Willcox
A.L. Eclipse Coordinator
Route 2, Box 940
Bartlesville, OK 74006

The A.L., a federation of astronomy clubs and societies, coordinates eclipse expeditions. The A.L. distributes a quarterly publication, *Reflector*, to its members, publishes a series of *Observe* guides, offers discounts on books to its members, and holds an annual convention.

The American Association of Variable Star Observers (A.A.V.S.O.)

25 Birch St.
Cambridge, MA 02138–1205

Founded in 1911, A.A.V.S.O. membership includes the twice–yearly *Journal of the American Association of Variable Star Observers*, and for nominal charges, *AAVSO Newsletter*, predictions of minima for variable stars, including eclipsing binaries, a catalog of available charts for observing variable stars, and individual star charts. Has a solar section specializing in sunspot counts. Holds meetings twice a year, in the spring and autumn.

International Occultation Timing Association (I.O.T.A.)

2760 S. W. Jewell Ave.
Topeka, KS 66611–1614

Membership includes local–site predictions of grazing occultations of bright stars by the moon, planets and asteroids, explanatory materials, and a subscription to the quarterly *Occultation Newsletter* and supplements. Total occultation predictions and detailed information on specific grazing occultations are available for nominal fees. Holds an annual meeting. Founded in 1982.

The American Meteor Society, Ltd. (A.M.S.)

Department of Physics and Astronomy
SUNY Geneseo
Geneseo, NY 14454–1401

The A.M.S. was founded in 1911. Membership includes the independent newsletter, *Meteor News*, as well as an annual report of society activities and member observations. Visual, telescopic, radio and spectroscopic observations, as well as photography, of meteor activity are encouraged.

The following publications are interested in receiving eclipse observations and photographs from amateurs; if unused material is to be returned, a self–addressed, stamped envelope should be provided.

Astronomy
Kalmback Publishing Co.
21027 Crossroads Circle
P. O. Box 1612
Waukesha, WI 53187

Sky & Telescope
CCD Astronomy*
P. O. Box 9111
Belmont, MA 02178–9111
(*CCD eclipse images only)

Meteor News**
Route 3, Box 1062
Callahan, FL 32011
(**Meteor reports only)

Appendix 5
List of Suppliers

> **NOTE:** No product or dealer recommendation is implied either by the authors or by the Astronomical League. Prospective buyers are urged to consult current issues of **Sky & Telescope**, **Astronomy** or **CCD Astronomy** magazines for changes, additions, and deletions to the list and current prices.

Solar Filters: The following suppliers or manufacturers advertise front–mounted solar filters and cells or solar viewers of various types for most popular makes of telescopes. Two advertise rear–mounted lunar filters.

Adorama
42 West 18th. Street
New York, NY 10011
(212) 741-0052
Stocks Thousand Oaks Optical glass filters.

J. M. B. Inc.
20762 Richard
Trenton, MI 48183
(313) 675-3490
Produces Identi–View™ glass solar filters of several types and custom sized to fit most telescopes and binoculars.

Meade Instruments Corporation
16542 Millikan Avenue
Irvine, CA 92714
(714) 756-1450
Moon filters; markets through authorized dealers.

Orion Telescope Center
2450 17th. Ave., P.O. Box 1158-S
Santa Cruz, CA 95061-1158
(408) 464-0446
Carries glass solar filters and rear–mounted lunar filters.

Pocono Mountain Optics
R.R. 6, Box 6329
Moscow, PA 18444
(717) 842-1500
Stocks Thousand Oaks Optical glass filters.

Rainbow Symphony, Inc.
6860 Canby Ave., #120
Reseda, CA 91335
(510) 581-8266
Carries Eclipse Shades™ mylar solar viewing glasses in cardboard frames for naked–eye viewing.

Sun Spotter
RD 1 Box 160
Hawley, PA 18428
(717) 685-7033
Produces a simple solar optical viewer useful for solar eclipses.

Thousand Oaks Optical
Box 5044-289
Thousand Oaks, CA 91359
(805) 491-3642
Produces several types of glass filters for telescopes, cameras, binoculars, and visual filters.

Roger W. Tuthill, Inc.
11 Tanglewood Lane
Mountainside, NJ 07092
(908) 232-1786
Produces Solar Skreen® mylar filters and cells for telescopes, binoculars, cameras and visual.

Electronic Imaging: The following suppliers or manufacturers offer CCD cameras and accessories; one offers photoelectric photometers.

CCD Technology
18092 Sky Park South, Unit E.
Irvine, CA 92714
(714) 752-2442
CCD-10 entry level imager.

CompuScope
3463 State Street, Suite 431
Santa Barbara, CA 93105
(805) 966-7179
CCD800/1600 series (integrating) cameras.

ELECTRonicIMaging Corporation
356 Wall St.
Princeton, NJ 08540
(609) 683-5546
EDC-1000 ISA Bus CCD cameras.

Murnaghan Instruments
1781 Primrose Ln.
West Palm Beach, FL 33414
(407) 795-2201
Imaging systems and accessories.

Optec, Inc.
199 Smith St.
Lowell, MI 49331
(616) 897-9351
SSP-5 PMT Stellar Photometers.

Santa Barbara Instrument Group
1482 East Valley Road, Suite #J601
Santa Barbara, CA 93108
(805) 969-1851
ST-4 Star Tracker/Imaging Camera.

Sirius Instruments Co.
141 N. Charles Ave.
Villa Park, IL 60181
(708) 782-5819
CWIP digital camera.

SpectraSource Instruments
31324 Via Colinas, Suite 114
Westlake Village, CA 91362
(818) 707-2655
Lynxx CCD cameras.

University Optics, Inc.
P. O. Box 1205
Ann Arbor, MI 48106
(313) 665-3575
Components for CCD Camera Cookbook kit.*

*Manual, image processing software and circuit boards for CCD Camera Cookbook available separately from **Willmann–Bell, Inc.**, P.O.Box 35025, Richmond, VA 23235, (804) 320-7016.

Appendix 6

Solar Eclipses, 1900-2025

1900–1935

Type of Eclipse: t=total, a=annular, a–t=annular–total, p=partial;
Path of Visibility: (centerline for t, a, a–t; entire region for p): 1=North America, 2=South America, 3=Europe, 4=Africa, 5=Asia, 6=Australia, 7=Antarctica, 8=Indonesia, 9=Greenland, 10=North Atlantic, 11=South Atlantic, 12=North Pacific, 13=South Pacific, 14=Indian, 15=Arctic, 16=Mediterranean Sea.

Year	Month	Day	Type	Path of Visibility	Saros #	Year	Month	Day	Type	Path of Visibility	Saros #
1900	5	28	t	1,10,3,16,4	126	1918	6	8	t	12,1,10	126
	11	22	a	4,14,6	131		12	3	a	13,2,11,4	131
1901	5	18	t	14,8	136	1919	5	29	t	13,2,11,10,4	136
	11	11	a	16,5,14,12	141		11	22	a	1,2,10,4	141
1902	4	8	p	15,1	108	1920	5	18	p	14,7,6	146
	5	7	p	13	146		11	10	p	1,9,10,3,16,4	151
	10	31	p	15,3,5	151	1921	4	8	a	10,15	118
1903	3	29	a	15,5	118		10	1	t	7,13	123
	9	21	t	14,7	123	1922	3	28	a	2,10,4	128
1904	3	17	a	4,14,5	128		9	21	t	4,14,6,13	133
	9	9	t	12,13,2	133	1923	3	17	a	11,4,14	138
1905	3	6	a	13,6	138		9	10	t	12,1,10	143
	8	30	t	1,10,16,5	143	1924	3	5	p	13,11,7,4	148
1906	2	23	p	7,14,6,13	148		7	31	p	13	115
	7	21	p	2,11,7	115		8	30	p	15,9,5	153
	8	20	p	1,9,15,5	153	1925	1	24	t	1,10	120
1907	1	14	t	5	120		7	20	a	13	125
	7	10	a	13,2,11	125	1926	1	14	t	4,14,8,12	130
1908	1	3	t	12,13,12,1	130		7	9	a	12	135
	6	28	a	12,1,10,4	135	1927	1	3	a–t	13,2,11	140
	12	23	a–t	13,2,11	140		6	29	t	1,3,5,15,12	145
1909	6	17	t	9,15,5	145		12	24	p	7,13,11	150
	12	12	p	7,11,13,6	150	1928	5	19	t	7,2,11,4	117
1910	5	9	t	6,7	117		6	17	p	15,5	155
	11	2	p	5,12,1	122		11	12	p	10,3,4,5,16,14	122
1911	4	28	t	13,12	127	1929	5	9	t	14,8,12	127
	10	22	a	5,12,13	132		11	1	a	10,4,14	132
1912	4	17	a–t	2,10,3,5	137	1930	4	28	a–t	12,1,10	137
	10	10	t	12,13,2,11	142		10	21	t	12,13	142
1913	4	6	p	12,1,9,15,5,12	147	1931	4	18	p	15,1,9,5	147
	8	31	p	1,9,10	114		9	12	p	12,1	114
	9	30	p	4,7,14	152		10	11	p	13,2,11,7	152
1914	2	25	a	7,13	119	1932	3	7	a	7	119
	8	21	t	15,3,5	124		8	31	t	15,1,10	124
1915	2	14	a	13,6,12	129	1933	2	24	a	11,4,14	129
	8	10	a	12,13	134		8	21	a	4,5,14,8,6,13	134
1916	2	3	t	12,2,10	139	1934	2	14	t	8,12	139
	7	30	a	14,6,13	144		8	10	a	11,4,14	144
	12	24	p	7,14	111	1935	1	5	p	(miss/graze) 13	111
1917	1	23	p	4,16,3,5,14	149		2	3	p	12,1,10,9	149
	6	19	p	1,15,9,5	116		6	30	p	15,1,9,10,3,5	116
	7	19	p	14,7	154		7	30	p	11,7	154
	12	14	a	7,13	121		12	25	a	7	121

Appendix 6, continued 1936–2025

Type of Eclipse: t=total, a=annular, a–t=annular–total, p=partial;
Path of Visibility: (centerline for t, a, a–t; entire region for p): 1=North America, 2=South America, 3=Europe, 4=Africa, 5=Asia, 6=Australia, 7=Antarctica, 8=Indonesia, 9=Greenland, 10=North Atlantic, 11=South Atlantic, 12=North Pacific, 13=South Pacific, 14=Indian, 15=Arctic, 16=Mediterranean Sea.

Year	Month	Day	Type	Path of Visibility	Saros #	Year	Month	Day	Type	Path of Visibility	Saros #
1936	6	19	t	16,5,12	126	1959	4	8	a	14,6,13	138
	12	13	a	6,13	131		10	2	t	1,10,4,14	143
1937	6	8	t	13,12,2	136	1960	3	27	p	14,7,6	148
	12	2	a	12	141		9	20	p	12,1,9,15,5	153
1938	5	29	t	11	146	1961	2	15	t	10,3,5	120
	11	21	p	5,12,1	151		8	11	a	7,11	125
1939	4	19	a	12,1,15	118	1962	2	5	t	8,13,12	130
	10	12	t	7	123		7	31	a	2,10,4	135
1940	4	7	a	13,12,1,10	128	1963	1	25	a	14,13,2,11,4	140
	10	1	t	12,2,11,4	133		7	20	t	12,1,10	145
1941	3	27	a	13,2	138	1964	1	14	p	13,11,7,2	150
	9	21	t	5,12	143		6	10	p	16,6,13,7	117
1942	3	16	p	13,7	148		7	9	p	1,15,9,5	155
	8	12	p	14,7	115		12	4	p	5,12,1	122
	9	10	p	1,9,15,3,4,5,16	153	1965	5	30	t	13	127
1943	2	4	t	5,12,1	120		11	23	a	5,8,12	132
	8	1	a	14	125	1966	5	20	a–t	10,4,16,5	137
1944	1	25	t	12,13,2,11,10,4	130		11	12	t	2,11,14	142
	7	20	a	4,14,5,12,13	135	1967	5	9	p	12,1,15,9,10,5	147
1945	1	14	a	13	140		11	2	t	7,11,14,4	152
	7	9	t	1,9,10,5	145	1968	3	28	p	13,7,2	119
1946	1	3	p	13,7,11,14	150		9	22	t	15,5	124
	5	30	p	13,2	117	1969	3	18	a	14,8,13,12	129
	6	29	p	1,15,9,3	155		9	11	a	12,13,2	134
	11	23	p	1,2,12,10,9	122	1970	3	7	t	13,12,1,10	139
1947	5	20	t	13,2,11,10,4	127		8	31	a	8,13	144
	11	12	a	12,13,2	132	1971	2	25	p	10,9,3,4,5,15	149
1948	5	9	a–t	14,5,12	137		7	22	p	5,15,1	116
	11	1	t	4,14	142		8	20	p	6,13,7	154
1949	4	28	p	1,15,9,10,4,16,5	147	1972	1	16	a	7,13	121
	10	21	p	13,7,6	152		7	10	t	5,12,1,10	126
1950	3	18	a	7	119	1973	1	4	a	13,2,11	131
	9	12	t	15,5,12	124		6	30	t	2,10,4,14	136
1951	3	7	a	13,12,2,10	129		12	24	a	12,2,10,4	141
	9	1	a	1,10,4	134	1974	6	20	t	14,6	146
1952	2	25	t	10,4,5	139		12	13	p	1,2,3,10,9	151
	8	20	a	13,2,11	144	1975	5	11	p	1,9,15,10,4,3,5	118
1953	2	14	p	5,12,1	149		11	3	p	7,2,13,11	123
	7	11	p	1,9,15	116	1976	4	29	a	1,4,16,5	128
	8	9	p	13,2,7,11	154		10	23	t	4,14,6,13	133
1954	1	5	a	7,13	121	1977	4	18	a	11,4,14	138
	6	30	t	1,9,10,5	126		10	12	t	12,2	143
	12	25	a	11,4,14	131	1978	4	7	p	13,7,2,11,4	148
1955	6	20	t	14,5,12,13	136		10	2	p	15,3,5,12	153
	12	14	a	4,14,5,12	141	1979	2	26	t	12,1,9	120
1956	6	8	t	13	146		8	22	a	13,7	125
	12	2	p	3,5,16,14	151	1980	2	16	t	11,4,14,5	130
1957	4	29	a	5,15	118		8	10	a	12,13,2	135
	10	23	p	13,7,11,4,14	123	1981	2	4	a	13	140
1958	4	19	a	14,5,12	128		7	31	t	5,12	145
	10	12	t	13	133	1982	1	25	p	13,7,11	150

Year	Month	Day	Type	Path of Visibility	Saros #	Year	Month	Day	Type	Path of Visibility	Saros #
	6	21	p	11,4,14	117	2004	4	19	p	11,7,4,14	119
	7	20	p	1,15,9,3,5	155		10	14	p	5,12,1,15	124
	12	15	p	3,4,5,16	122	2005	4	8	a–t	13,12,2	129
1983	6	11	t	14,8,13	127		10	3	a	10,3,16,4,14	134
	12	4	a	10,11,4	132	2006	3	29	t	2,11,10,4,16	139
1984	5	30	a–t	12,1,10,4	137		9	22	a	2,10,11,14	144
	11	22	t	8,13	142	2007	3	19	p	14,5,15,1	149
1985	5	19	p	12,1,15,9,10,5	147		9	11	p	13,2,11,7	154
	11	12	t	13	152	2008	2	7	a	7,13	121
1986	4	9	p	7,13,14,6	119		8	1	t	1,9,15,5	126
	10	3	t	10	124	2009	1	26	a	11,14,8	131
1987	3	29	a–t	2,13,4,14	129		7	22	t	14,5,12,13	136
	9	23	a	5,12,13	134	2010	1	15	a	4,14,5	141
1988	3	18	t	14,8,12	139		7	11	t	13,2	146
	9	11	a	14	144	2011	1	4	p	10,3,4,5,16	151
1989	3	7	p	12,1,15,9	149		6	1	p	15,1,10,9,5,12	118
	8	31	p	4,14,7	154		7	1	p	11,7,14	156
1990	1	26	a	7,11	121		11	25	p	13,7,11,4,14	123
	7	22	t	5,15,12	126	2012	5	20	a	5,12,1	128
1991	1	15	a	14,6,13	131		11	13	t	6,13	133
	7	11	t	12,1,2	136	2013	5	10	a	6,8,13,12,13	138
1992	1	4	a	12,13,12,1	141		11	3	t	10,4	143
	6	30	t	2,11,14	146	2014	4	29	a	7	148
	12	24	p	5,12,1	151		10	23	p	12,1,15,5	153
1993	5	21	p	12,1,15,9,10,3,5	118	2015	3	20	t	10,15	120
	11	13	p	13,2,11,7,6	123		9	13	p	11,4,14,7	125
1994	5	10	a	12,1,10,4	128	2016	3	9	t	14,8,12	130
	11	3	t	13,2,11,14	133		9	1	a	11,4,14	135
1995	4	29	a	13,2,11	138	2017	2	26	a	13,2,11,4	140
	10	24	t	5,8,12	143		8	21	t	12,1,10	145
1996	4	17	p	13,7	148	2018	2	15	p	7,13,11,2	150
	10	12	p	1,10,9,3,4,16	153		7	13	p	14,6,7,13	117
1997	3	9	t	5,15	120		8	11	p	15,1,9,10,5	155
	9	2	p	6,7,13	125	2019	1	6	p	5,12,1	122
1998	2	26	t	13,12,2,10	130		7	2	t	13,2	127
	8	22	a	14,8,13	135		12	26	a	5,14,8,12	132
1999	2	16	a	11,14,6,13	140	2020	6	21	a	4,5,12	137
	8	11	t	10,3,5,14	145		12	14	t	13,2,11	142
2000	2	5	p	13,11,7,14	150	2021	6	10	a	1,9,15,5	147
	7	1	p	13,2	117		12	4	t	11,7,13	152
	7	31	p	12,1,15,9,5	155	2022	4	30	p	13,7,2,11	119
	12	25	p	12,1,10,9	122		10	25	p	9,10,3,4,16,5,14	124
2001	6	21	t	11,4,14	127	2023	4	20	t	14,6,8,12	129
	12	14	a	12,13,12,11	132		10	14	a	13,12,1,2,11	134
2002	6	10	a	8,12,1	137	2024	4	8	t	13,12,1,10	139
	12	4	t	2,4,14,6	142		10	2	a	12,13,2,11	144
2003	5	31	a	9,10	147	2025	3	29	p	1,15,9,10,3,4,5	149
	11	23	t	14,7	152		9	21	p	13,7	154

Total Solar Eclipses: 1994 - 2020

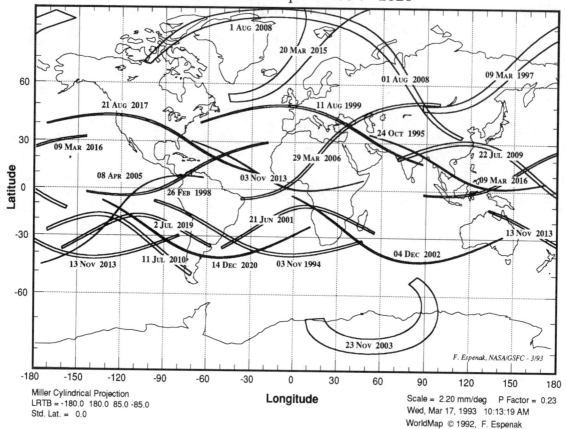

Annular Solar Eclipses: 1994 - 2020

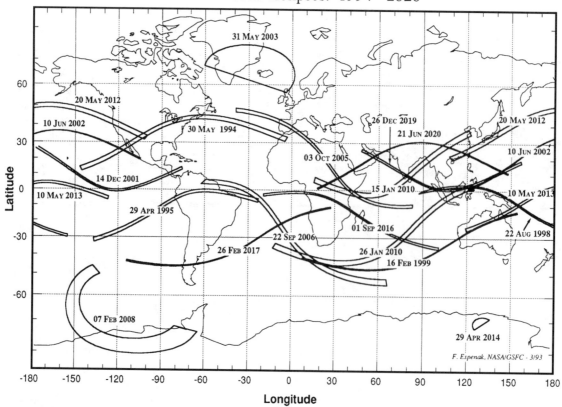

Lunar Eclipses, 1900-2025

1900–1934

Day: (UT of maximum phase);
Type of Eclipse: t=total, p=partial, n=penumbral;
Magnitude: For penumbral eclipses, a magnitude of 0.500 is considered minimum for visibility;
Visibility Region: (maximum phase for t; whole region for p, n): 1=E. Australia, Pacific, W. North America; 2=Pacific, N. America, W. South America; 3=E. North America, S. America, W. Atlantic; 4=E. South America, E. Atlantic, W. Africa, W. Europe; 5=E. Africa, E. Europe, W. Asia; 6=E. Africa, E. Europe, central Asia, Indian Ocean, W. Australia; 7=E. Asia, Australia, Pacific.

Year	Month	Day	Type	Magnitude	Path of Visibility	Saros #	Year	Month	Day	Type	Magnitude	Path of Visibility	Saros #
1900	6	13	p	0.001	3–4	167		7	4	t	1.623	5	155
	12	6	n	0.844	7–2	173		12	28	t	1.009	1–2	161
1901	5	3	n	1.069	6	178	1918	6	24	p	0.134	1	167
	10	27	p	0.227	7	184		12	17	n	0.859	4–6	173
1902	4	22	t	1.337	6	190	1919	5	14	n	0.937	4	178
	10	17	t	1.463	2	196		11	7	p	0.183	3–5	184
1903	4	12	p	0.973	4	202	1920	5	3	t	1.224	4	190
	10	6	p	0.869	7	208		10	27	t	1.405	7	196
1904	3	2	n	0.201	2–4	213	1921	4	22	t	1.072	2	202
	3	31	n	0.729	7–1	214		10	16	p	0.936	4–5	208
	9	24	n	0.568	5–7	220	1922	3	13	n	0.159	7–1	213
1905	2	19	p	0.412	6	2		4	11	n	0.806	6	214
	8	15	p	0.291	3	8		10	6	n	0.660	3–5	220
1906	2	9	t	1.631	2	14	1923	3	3	p	0.376	3–4	2
	8	4	t	1.785	7	20		8	26	p	0.167	1	8
1907	1	29	p	0.714	7	26	1924	2	20	t	1.604	7	14
	7	25	p	0.622	3	32		8	14	t	1.658	6	20
1908	1	18	n	0.562	7–1	38	1925	2	8	p	0.734	4–5	26
	6	14	n	0.839	7	43		8	4	p	0.753	1	32
	7	13	n	0.254	4–5	44	1926	1	28	n	0.580	4–5	38
	12	7	n	1.060	4–5	49		6	25	n	0.699	4–5	43
1909	6	4	t	1.163	4	55		7	25	n	0.380	2–3	44
	11	27	t	1.371	2	61		12	19	n	1.052	2	49
1910	5	24	t	1.099	2–3	67	1927	6	15	t	1.016	1	55
	11	17	t	1.132	3–5	73		12	8	t	1.356	6	61
1911	5	13	n	0.825	2–3	79	1928	6	3	t	1.247	7–1	67
	11	6	n	0.842	6–7	85		11	27	t	1.155	2	73
1912	4	1	p	0.187	4–5	90	1929	5	23	n	0.962	7	79
	9	26	p	0.123	7–1	96		11	17	n	0.872	3–5	85
1913	3	22	t	1.574	7–1	102	1930	4	13	p	0.113	2–3	90
	9	15	t	1.435	7–1	108		10	7	p	0.029	6	96
1914	3	12	p	0.916	3	114	1931	4	2	t	1.509	5	102
	9	4	p	0.862	7	120		9	26	t	1.325	5	108
1915	1	31	n	0.071	2–4	125	1932	3	22	p	0.971	7–1	114
	3	1	n	0.580	6	126		9	14	p	0.979	5	120
	7	26	n	0.379	7	131	1933	2	10	n	0.044	7–1	125
	8	24	n	0.600	4–5	132		3	12	n	0.617	3–4	126
1916	1	20	p	0.137	2	137		8	5	n	0.259	6	131
	7	15	p	0.800	3	143		9	4	n	0.719	2–3	132
1917	1	8	t	1.368	2	149	1934	1	30	p	0.115	7	137

Appendix 7, continued 1934-2025

Day: (UT of maximum phase);
Type of Eclipse: t=total, p=partial, n=penumbral;
Magnitude: For penumbral eclipses, a magnitude of 0.500 is considered minimum for visibility;
Visibility Region: (maximum phase for t; whole region for p, n): 1=E. Australia, Pacific, W. North America; 2=Pacific, N. America, W. South America; 3=E. North America, S. America, W. Atlantic; 4=E. South America, E. Atlantic, W. Africa, W. Europe; 5=E. Africa, E. Europe, W. Asia; 6=E. Africa, E. Europe, central Asia, Indian Ocean, W. Australia; 7=E. Asia, Australia, Pacific.

Year	Month	Day	Type	Magnitude	Path of Visibility	Saros #
	7	26	p	0.667	7	143
1935	1	19	t	1.354	7	149
	7	16	t	1.759	3	155
1936	1	8	t	1.022	6–7	161
	7	4	p	0.271	7	167
	12	28	n	0.870	2–4	173
1937	5	25	n	0.796	2	178
	11	18	p	0.150	2	184
1938	5	14	t	1.101	1	190
	11	7	t	1.359	4–5	196
1939	5	3	t	1.183	7	202
	10	28	p	0.991	2	208
1940	3	23	n	0.105	4–6	213
	4	22	n	0.894	2–3	214
	10	16	n	0.740	2	220
1941	3	13	p	0.330	1	2
	9	5	p	0.055	6–7	8
1942	3	3	t	1.568	4–5	14
	8	26	t	1.539	3	20
1943	2	20	p	0.763	2–3	26
	8	15	p	0.877	6	32
1944	2	9	n	0.605	2–4	38
	7	6	n	0.559	2–3	43
	8	4	n	0.503	7	44
	12	29	n	1.047	7	49
1945	6	25	p	0.864	7	55
	12	19	t	1.347	3–4	61
1946	6	14	t	1.402	6	67
	12	8	t	1.170	6	73
1947	6	3	p	0.025	6	79
	11	28	n	0.894	2	85
1948	4	23	p	0.028	7	90
	10	18	n	1.040	3–4	96
1949	4	13	t	1.432	3	102
	10	7	t	1.229	3–4	108
1950	4	2	t	1.038	5	114
	9	26	t	1.083	3	120
1951	2	21	n	0.007	5–6	125
	3	23	n	0.667	7–1	126
	8	17	n	0.145	2–4	131
	9	15	n	0.828	7–1	132
1952	2	11	p	0.087	3–5	137
	8	5	p	0.538	6	143
1953	1	29	t	1.336	4–5	149
	7	26	t	1.870	7	155
1954	1	19	t	1.035	3–4	161
	7	16	p	0.409	3–4	167
1955	1	8	n	0.880	7–1	173

Year	Month	Day	Type	Magnitude	Path of Visibility	Saros #
	6	5	n	0.648	7	178
	11	29	p	0.125	6–7	184
1956	5	24	p	0.969	7	190
	11	18	t	1.323	2	196
1957	5	13	t	1.303	4–5	202
	11	7	t	1.034	7	208
1958	4	4	n	0.040	2–4	213
	5	3	p	0.015	7	214
	10	27	n	0.807	7	220
1959	3	24	p	0.271	5	2
	9	17	n	1.012	3–4	8
1960	3	13	t	1.519	2	14
	9	5	t	1.430	7–1	20
1961	3	2	p	0.804	7	26
	8	26	p	0.993	3–4	32
1962	2	19	n	0.638	7–1	38
	7	17	n	0.418	7	43
	8	15	n	0.621	6	44
1963	1	9	n	1.044	4–5	49
	7	6	p	0.710	4–5	55
	12	30	t	1.341	7–1	61
1964	6	25	t	1.561	3–4	67
	12	19	t	1.181	3–4	73
1965	6	14	p	0.180	3–4	79
	12	8	n	0.907	6	85
1966	5	4	n	0.942	5	90
	10	29	n	0.977	1	96
1967	4	24	t	1.343	7	102
	10	18	t	1.146	1	108
1968	4	13	t	1.117	2–3	114
	10	6	t	1.173	1	120
1969	4	2	n	0.728	6	126
	8	27	n	0.039	7–2	131
	9	25	n	0.925	4–5	132
1970	2	21	p	0.050	2	137
	8	17	p	0.414	3–4	143
1971	2	10	t	1.312	2	149
	8	6	t	1.735	6	155
1972	1	30	t	1.054	1	161
	7	26	p	0.547	2	167
1973	1	18	n	0.891	4–5	173
	6	15	n	0.495	5–6	178
	7	15	n	0.130	7–1	179
	12	10	p	0.107	3–4	184
1974	6	4	p	0.831	5	190
	11	29	t	1.296	7	196
1975	5	25	t	1.431	2–3	202
	11	18	t	1.067	4–5	208

Year	Month	Day	Type	Magnitude	Path of Visibility	Saros #
1976	5	13	p	0.127	5–6	214
	11	6	n	0.862	4–5	220
1977	4	4	p	0.201	2–3	2
	9	27	n	0.925	2	8
1978	3	24	t	1.459	7	14
	9	16	t	1.331	5–6	20
1979	3	13	p	0.857	5	26
	9	6	t	1.101	1	32
1980	3	1	n	0.680	4–5	38
	7	27	n	0.280	5–7	43
	8	26	n	0.733	3–4	44
1981	1	20	n	1.039	2	49
	7	17	p	0.555	3	55
1982	1	9	t	1.336	4–5	61
	7	6	t	1.722	1–2	67
	12	30	t	1.188	7–1	73
1983	6	25	p	0.339	1–2	79
	12	20	n	0.915	3–4	85
1984	5	15	n	0.832	3	90
	6	13	n	0.090	7	91
	11	8	n	0.926	5–6	96
1985	5	4	t	1.243	5	102
	10	28	t	1.078	5–6	108
1986	4	24	t	1.207	7	114
	10	17	t	1.249	5	120
1987	4	14	n	0.803	3–4	126
	10	7	n	1.010	2–4	132
1988	3	3	p	0.002	7	137
	8	27	p	0.297	1	143
1989	2	20	t	1.279	7	149
	8	17	t	1.605	3	155
1990	2	9	t	1.078	6	161
	8	6	p	0.682	7	167
1991	1	30	n	0.905	2–3	173
	6	27	n	0.339	3	178
	7	26	n	0.279	5–7	179
	12	21	p	0.094	1	184
1992	6	15	p	0.687	3	190
	12	9	t	1.276	3–5	196
1993	6	4	t	1.566	7	202
	11	29	t	1.092	2	208
1994	5	25	p	0.248	3	214
	11	18	n	0.906	2	220
1995	4	15	p	0.118	7	2
	10	8	n	0.850	6	8
1996	4	4	t	1.385	4–5	14
	9	27	t	1.245	3–4	20
1997	3	24	p	0.923	3	26
	9	16	t	1.197	5–6	32
1998	3	13	n	0.734	2–4	38
	8	8	n	0.146	3–4	43
	9	6	n	0.837	7–1	44
1999	1	31	n	1.030	6–7	49
	7	28	p	0.401	1	55
2000	1	21	t	1.331	2–4	61
	7	16	t	1.773	7	67
2001	1	9	t	1.194	4–6	73
	7	5	p	0.499	7	79
	12	30	n	0.918	7–2	85
2002	5	26	n	0.715	7	90
	6	24	n	0.234	4–5	91
	11	20	n	0.886	3–4	96
2003	5	16	t	1.134	3	102
	11	9	t	1.022	3–4	108
2004	5	4	t	1.309	5	114
	10	28	t	1.312	3–4	120
2005	4	24	n	0.890	1	126
	10	17	p	0.067	7–1	132
2006	3	14	n	1.056	3–5	137
	9	7	p	0.189	5–6	143
2007	3	3	t	1.237	4–5	149
	8	28	t	1.482	1	155
2008	2	21	t	1.111	3–4	161
	8	16	p	0.812	5	167
2009	2	9	n	0.925	7	173
	7	7	n	0.182	1–2	178
	8	6	n	0.427	3–4	179
	12	31	p	0.083	4–5	184
2010	6	26	p	0.541	1	190
	12	21	t	1.262	2	196
2011	6	15	t	1.706	6	202
	12	10	t	1.109	7	208
2012	6	4	p	0.376	1	214
	11	28	n	0.940	7–1	220
2013	4	25	p	0.023	6	2
	5	25	n	0.038	1–4	3
	10	18	n	0.790	3–5	8
2014	4	15	t	1.297	2	14
	10	8	t	1.171	1	20
2015	4	4	t	1.003	7–1	26
	9	28	t	1.283	3–4	32
2016	3	23	n	0.800	7–1	38
	8	18	n	0.019	1–2	43
	9	16	n	0.932	6	44
2017	2	11	n	1.014	3–5	49
	8	7	p	0.252	6	55
2018	1	31	t	1.321	7	61
	7	27	t	1.614	6	67
2019	1	21	t	1.201	2–3	73
	7	16	p	0.658	5	79
2020	1	10	n	0.921	6	85
	6	5	n	0.593	6	90
	7	5	n	0.380	3	91
	11	30	n	0.855	1–2	96
2021	5	26	t	1.016	1	102
	11	19	p	0.979	1–2	108
2022	5	16	t	1.420	3	114
	11	8	t	1.363	1	120
2023	5	5	n	0.989	6–7	126
	10	28	p	0.127	4–5	132
2024	3	25	n	0.981	1–2	137
	9	18	p	0.091	3–4	143
2025	3	14	t	1.183	2	149
	9	7	t	1.367	6	155

Appendix 8

Astronomical League Solar Eclipse Report Form

Part I: Observer/Site Information
Name: (Last)   (First)   (MI)
Address:
City: State: Zip: Phone:( ) –
Date of Eclipse (UT):   Date of Report:
Type of Eclipse At Your Site (circle):   Lunar / Solar   Penumbral / Partial / Total / Annular
Expedition Group Name (if any):
Observing Site Location:
Site Latitude: Longitude: Elevation:
Site Description (routes, roads, airport, etc.):

Part II: Weather/Seeing/Transparency Conditions
Wind direction (from):   Wind velocity: Humidity: %
Barometric pressure: Cloud type(s):   Cloud cover: %
Seeing conditions (0–10):   Atmospheric transparency (0–5):
Air temperature–1st. Contact: 3rd. Contact: 4th. Contact:
Notes:

Part III: Eclipse Circumstances
Lunar/Solar position at totality – Azimuth: Altitude:
Timings (UT):
     1st. contact: 3rd. contact:
     2nd. contact: 4th. contact:
Duration of totality in minutes and seconds:

Part IV: Experiments
Title and purpose:
Description:

Equipment:

Participating observers:

Research references:

Summarize data and results on additional sheets and attach.

Eclipse Observer's Log

Year	Mo.	Day	Type[1]	Observing site[2]	Outcome[3]	Duration[4]

[1] Type of eclipse: S=solar, L=lunar; p=partial, a=annular, a–t=annular–total, t=total, n=penumbral.
[2] Observing site: Latitude & longitude or place name.   [3] Degree of success, remarks.   [4] Duration of totality at site.

Appendix 10

A.L.P.O. Galilean Satellite Eclipse Visual Timing Report Form

| Describe your time source(s) and estimated accuracy: | Observer Name: |
| | Apparition: 19___ - 19___ (conjunction to conjunction) |

Event Type (a)	Predicted UT		Observed UT Time (d)	Telescope Data			Sky Conditions (0-2 scale) (f)			Notes (g)
	Date (b)	Time (c)		Type (e)	Aperture (cm)	Magnification	Seeing	Transparency	Field Brightness	

Notes

(a) 1 = Io, 2 = Europa, 3 = Ganymede, 4 = Callisto; D = Disappearance, R = Reappearance.   (b) Month and Day.   (c) To 1 minute.

(d) To 1 second, corrected for watch error if applicable. Indicate in "Notes" if the observed UT date differs from the predicted UT date.

(e) R = Refractor, N = Newtonian Reflector, C = Cassegrain Reflector, X = Compound/Catadioptric System; if other type, indicate in "Notes."

(f) These conditions, including field brightness (due to moonlight, twilight, etc.), should be described as they apply to the actual field of view, rather than general sky conditions. Use whole numbers only, as follows:
    0 = Condition not perceptible; no effect on timing accuracy.
    1 = Condition perceptible; possible minor effect on timing accuracy.
    2 = Condition serious; definite effect on timing accuracy.

(g) Note here such factors as wind, drifting cloud, satellite near Jupiter's limb, moonlight interference, and so forth.

Send reports to: Dr. John E. Westfall, P.O. Box 16131, San Francisco CA 94116

Appendix 11

Request Form for NASA Solar Eclipse Bulletins

NASA Solar Eclipse Bulletins contain detailed predictions, maps and meteorology for future central solar eclipses of interest. Part of NASA's Reference Publication (RP) series, the Bulletins are provided as a service to the international astronomical community in the planning and execution of successful eclipse expeditions. They should also be useful to educators, the media and the lay community as a source of concise eclipse information. Comments, suggestions and corrections are solicited to improve subsequent Bulletins.

To allow lead time for planning purposes, NASA Solar Eclipse Bulletins are be published 18 to 24 months before each event. Single copies of the Bulletins are available at no cost. All requests must be accompanied by a 9" x 12" self–addressed stamped envelope (SASE) with sufficient postage for each Bulletin (11 oz. or 310 g.). Use stamps only since cash or checks can not be accepted. Please print either the RP–number or the eclipse date (year & month) of the Bulletin ordered in the lower left corner of the SASE and return with a copy of this form which has been completed to either of the below addresses. Requests from outside the United States and Canada may use International postal coupons to cover postage. Exceptions to the postage requirements will be made for international requests where political or economic restraints prevent the transfer of funds to other countries.

Permission is freely granted to reproduce any portion of the NASA Solar Eclipse Bulletins. All uses and/or publication of this material should be accompanied by an appropriate acknowledgment of the source.

Name of Organization: _____
(in English, if necessary): _____
Name of Contact Person: _____
Address: _____
City/State/ZIP: _____
Country: _____

Type of Organization: ___ University/College   ___ Observatory   ___ Library
(check all which apply)   ___ Planetarium   ___ Publication   ___ Media
                           ___ Professional   ___ Amateur   ___ Individual

Size of Organization: _____ (Number of Members)

Activities: _____
_____

a) ___Total Solar   ___Annular Solar of ____/____/____ (day/month/year)
b) ___Total Solar   ___Annular Solar of ____/____/____ (day/month/year)

Return Requests and Comments to:
    Fred Espenak   or   Jay Anderson
    NASA/GSFC   Environment Canada
    Code 693   900–266 Graham Avenue
    Greenbelt, MD 20771 USA   Winnipeg, MB, CANADA R3C 3V4
    Internet: u32fe@lepvax.gsfc.nasa.gov   Bitnet: jander@ccu.umanitoba.ca

Comments on and corrections to the NASA Solar Eclipse Bulletin Reference Publications can also be made to Jay Anderson or Fred Espenak via the above e–mail addresses.

Appendix 11, continued

NASA Solar Eclipse Bulletins on the Internet

Through the efforts of Dr. Joe Gurman (Goddard Space Flight Center/Solar Physics Branch), the Solar Eclipse Bulletins of the 10 May 1994 Annular Solar Eclipse and the 4 November 1994 Total Solar Eclipse are now available over the Internet. Formats include BinHex–encoded versions of the original Microsoft Word for Mac text files and those figures generated in PICT format, GIF versions of all the figures, and JPEG scans of the GNC maps of the eclipse path, as well as hypertext versions. They can be read or downloaded via the World–Wide Web server with a mosaic client from the Solar Data Analysis Center home page:

http://umbra.gsfc.nasa.gov/sdac.html

The top–level URL for the eclipse bulletins themselves are:

http://umbra.gsfc.nasa.gov/eclipse/940510/rp.html

and

http://umbra.gsfc.nasa.gov/eclipse/941103/rp.html

BinHex–encoded, StuffIt Lite–compressed version of the original Word and PICT files are available via anonymous ftp at:

file://umbra.gsfc.nasa.gov/pub/eclipse

Directories of GIF figures, ASCII tables, and JPEG maps are also accessible through the Web.

In addition to the NASA Solar Eclipse Bulletins, tables have been generated for all central solar eclipses from 1995 through 2000. The eclipse path predictions were generated using the JPL DE/LE 200 ephemeris using the center of mass for the moon. No corrections have been made to adjust for center of figure. The value used for the Moon's mean radius is k=0.272281. The umbral path characteristics have been predicted at 2 minute intervals of time compared to the 6 minute interval used in *Fifty Year Canon of Solar Eclipses*: 1986–2035. This should provide enough detail for making preliminary plots of the path on larger scale maps. Note that positive latitudes are north and positive longitudes are west.

The paths for the following 7 eclipses are now available via the Internet:

29 April 1995 – Annular Solar Eclipse
24 October 1995 – Total Solar Eclipse
9 March 1997 – Total Solar Eclipse
26 February 1998 – Total Solar Eclipse
22 August 1998 – Annular Solar Eclipse
16 February 1999 – Annular Solar Eclipse
11 August 1999 – Total Solar Eclipse

The tables can be accessed with Mosaic through SDAC home page, or directly at URL:

http://umbra.gsfc.nasa.gov/eclipse/predictions/eclipse-paths.html

All future NASA Solar Eclipse Bulletins will also be available over the Internet. Please report comments, corrections or suggestions to Fred Espenak ("u32fe@lepvax.gsfc.nasa.gov"). For Internet–related problems, please contact Joe Gurman ("gurman@uvsp.gsfc.nasa.gov").

Fred Espenak
NASA/Goddard SpaceFlight Center
Planetary Systems Branch, Code 693
Greenbelt, MD 20771 USA

Index

Agathocles, 2
American Association of Variable Star Observers, 53, 56
American Meteor Society, 53
Anderson, Jay, 5, 19
Archilochus, 3
archiving photographs, 65
Aristophanes, 3
artificial satellites
    observed during lunar eclipses, 53
    observed during solar eclipses, 39
ascending node, 10
Association of Lunar and Planetary Observers, 47–48, 49, 53, 56, 65
Assyrian eclipse records, 1
atmospheric extinction, 49

Babylonian eclipse predictions, 2
Baily, Francis, 2, 31
Baily's beads, 2, 31–33, 34
brightness (lunar), 41, 48–49
Butler, Howard Russell, 29

camera
    bodies, 60
    focusing, 60
    lens systems, 60
Chaldean astronomical almanac, 1
charge–coupled devices
    description, 67–68
    exposures, 68–69
    image coverage, 68
Chinese eclipse predictions, 1–2
chromosphere, 34, 35
Clark, George and Alvan G., 2
Clarke, Arthur C., 54
Columbus, Christopher, 2
comets
    observed during lunar eclipses, 53
    observed during solar eclipses, 39
commercial eclipse expeditions, 18
contact timings
    crater
        lunar, 40, 42, 49
        solar, 23, 31, 34
        sunspot, 23
        transit, 54
Cook, Captain James, 54
corona, 36–37
Crabtree, William, 54

Danjon, André–Louis, 52
Danjon Luminosity Scale, 52
descending node, 10
diamond ring, 31–33
Dunham, David, 32

eclipse
    artificial satellite events, 57
    central, 7
    duration, 11
    lunar, 7
        partial, 7–8
        penumbral, 7–8
        total, 7–8
    magnitude, 11
    planetary satellite, 55–56
    seasons, 10–11
    solar, 7
        annular, 8–9
        annular–total, 9–10
        partial, 9–10
        total, 7–8
    year, 10–11
eclipsing binary stars, 57
Edberg, Stephen, 5
Einstein's General Theory of Relativity, 38–39
Electromagnetic spectrum, 13
Espenak, Fred,
exposures; photographic
    lunar eclipses
        multiple exposures, 64–65
        partial, 50, 63–64
        penumbral, 41
        planets and stars, 64
        totality, 64
    solar eclipses
        annular, 61
        atmospheric effects, 63
        Baily's beads, 61–62
        chromosphere, 62
        corona, 37, 62
        diamond ring, 61–62
        multiple exposures, 24, 62
        partial, 24, 61
        planets and stars, 39, 63
        prominences, 62
        shadow bands, 62–63
        sunrise–sunset effect, 63
        totality, 61
exposures; video, 66

film
    black & white, 59
    color print, 59
    color slide, 59
filters
    color, 41, 51
    front–mounted, 15–16
    Herschel wedge, 15
    neutral density, 40, 42, 48

filters (cont.)
    polarizing, 40
    rear-mounted eyepiece, 15
    solar, 14–17
flash spectrum, 34

Garcia, David
Gassendi, Pierre, 54
gegenschein
    observed during solar eclipses, 39
General theory of relativity, 38–39
geoumbrascope, 49
Glenn, William, 29–30
Goldschmidt, H., 27
Graham, Francis, 49

Haggard, Sir Henry Rider, 3
Halley, Sir Edmund, 31
Hardy, Thomas, 3
Hines, C. F., 33
Homer, 3
Horrocks, Jeremiah, 54
Hsi and Ho, 1–2
Humphries, W. J., 21

Inex series, 12
intensity (shadow), 41, 48–49
International Occultation Timing Association, 32, 52, 53, 56

Kepler, Johannes, 54
King, Stephen, 4
Kidinnu, 2

Levy, David, 4
Lindsay, Lord, i
line of nodes, 10
Lomonosov's phenomenon, 54
lunar transient phenomena, 41, 49–50, 52

Meier, Ludwig, 5
meteors
    observed during lunar eclipses, 53
    observed during solar eclipses, 39
Meton, 2, 12
Metonic cycle, 12
Milton, 3

Nabu-rimannu, 2
Ney, Edward P., 39
node
    ascending, 10
    descending, 10
    line of, 10

occultations, 32, 56–57
    observed during lunar eclipses, 52–53
Olivarez, Jose, 5

Pasachoff, Jay, 4
Paulton, Edgar, 27–28
penumbra, 7
photographic photometry, 41, 49–50, 53
Pillmore, Dorothy, 4

Pindar, 3
Planets
    observed during solar eclipses, 39
Plutarch, 3, 36
projection method of solar viewing
    mirror, 14–15
    optical, 15, 24, 34
    pinhole, 14, 25
prominence, 34, 35
Ptolemy, 1

Reynolds, Mike, 6
Rittenhouse, David, 2
Roberts, Carter, 5
Royal Astronomical Society of Canada, 15, 48

Saros cycle, 12
shadow band screen, 27–28
shadow variations, 41, 48, 51
Sperling's Eight-Second Law, 33
Sperling, Norman, 28, 33
stars
    observed during solar eclipses, 39
streamers, 37
Sweetsir, Richard, 6
sunrise-sunset effect, 29, 30, 63

Thales, 2
Thucydides, 1, 2
time signals, 23, 47
transits, 54–55
    Mercury, 54
    Venus, 54–55
Trombino, Donald, 20

umbra, 7

van den Bergh, G., 12
variable stars
    observed during lunar eclipses, 53
    observed during solar eclipses, 39
VHS and S-VHS tapes, 66
video cameras
    batteries, 66–67
    controls, 66–67
    exposures, 68–69
    lens systems, 67
    mounting, 67
    systems in general, 66, 67
    telextenders, 67
vision safety, 6, 13–17, 33, 35, 37, 61

weather, 19, 25–26
wildlife, 25
Williams, Samuel, 2, 31
Wilson, H. C., 21
Winlock, Joseph, 3

Young, C. A., 34–35

Zodiacal light
    observed during solar eclipses, 39